Palgrave Studies in Climate Resilient Societies

Series Editor
Robert C. Brears
Avonhead, Canterbury, New Zealand

The Palgrave Studies in Climate Resilient Societies series provides readers with an understanding of what the terms **resilience and climate resilient** societies mean; the best practices and lessons learnt from various governments, in both non-OECD and OECD countries, implementing climate resilience policies (in other words what is 'desirable' or 'undesirable' when building climate resilient societies); an understanding of what a resilient society potentially looks like; knowledge of when resilience building requires slow transitions or rapid transformations; and knowledge on how governments can create coherent, forward-looking and flexible policy innovations to build climate resilient societies that: support the conservation of ecosystems; promote the sustainable use of natural resources; encourage sustainable practices and management systems; develop resilient and inclusive communities; ensure economic growth; and protect health and livelihoods from climatic extremes.

More information about this series at
http://www.palgrave.com/gp/series/15853

Robert C. Brears

Developing the Circular Water Economy

Robert C. Brears
Our Future Water
Christchurch, New Zealand

ISSN 2523-8124   ISSN 2523-8132   (electronic)
Palgrave Studies in Climate Resilient Societies
ISBN 978-3-030-32574-9   ISBN 978-3-030-32575-6   (eBook)
https://doi.org/10.1007/978-3-030-32575-6

This Palgrave Pivot imprint is published by the registered company Springer Nature
Switzerland AG
The registered company address is: Gewerbestrasse 11, 6330 Cham, Switzerland

ACKNOWLEDGEMENTS

First, I wish to first thank Rachael Ballard, who is not only a wonderful commissioning editor but a visionary who enables books like mine to come to fruition. Second, I wish to thank Mum, who has a great interest in the environment and water-related issues and has supported me in this journey of writing the book.

CONTENTS

LIST OF TABLES

CHAPTER 1

Introduction

Abstract This book discusses how the development of the circular water economy, which mitigates emissions and enhances resilience to climate change, can be guided by innovative policies that encourage the reducing of water consumption, reusing and recycling of water, and recovery of materials from wastewater.

Keywords Circular economy · Linear economy · Resource scarcity · Climate change

In our current economic model, manufactured capital, human capital, and natural capital all contribute to human welfare by supporting the production of goods and services in the economic process, where natural capital—the world's stock of natural resources (provided by nature before their extraction or processing by humans)—is typically used for material and energy inputs into production and acts as a 'sink' for waste from the economic process.[1] This economic model can be best described as 'linear' which typically involves economic actors[2]—who are people or organisations engaged in any of the four economic activities of production, distribution, consumption, and resource maintenance—harvesting and extracting natural resources, using them to manufacture a product, and selling a product to other economic actors, who then discard it when it no longer serves its purpose.[3]

© The Author(s) 2020

R. C. Brears, *Developing the Circular Water Economy*,
Palgrave Studies in Climate Resilient Societies,
https://Doi.org/10.1007/978-3-030-32575-6_1

1

In the linear economy, following the Take-Make-Dispose model, the water sector typically employs the Take-Use-Discharge strategy. In this strategy, water is 'withdrawn' from streams, rivers, lakes, reservoirs, oceans, and groundwater reservoirs as well as harvested directly as rainwater. Water is then 'used' by municipalities, industries, agriculture, the environment, etc. within the water cycle, including for consumptive and non-consumptive uses. Non-consumptive used water is 'returned' to the river basin directly or via a municipal treatment facility. Depending on the location within the basin this returned water could then be used downstream or lost to the basin. While the current linear economic model has generated an unprecedented level of growth, the model has led to constraints on the availability of water resources in addition to the generation of waste and environmental degradation from a variety of climatic and non-climatic trends.

In response to climate change, increasing resource scarcity, and environmental degradation, governments around the world are implementing a variety of policies to encourage the transition towards the 'circular economy' that focuses on reducing material consumption, reusing materials, and recovering materials from waste. In the context of water resources management, water utilities are beginning to promote the circular water economy that reduces water consumption, reuses and recycles water and wastewater, and recovers materials, including heat and minerals, from water and wastewater to not only mitigate greenhouse gas emissions but also enhance resilience to climate change from efficiency gained in reducing water consumption and reusing water for various activities.

There is however no systematic survey of the various policy solutions water utilities have implemented to encourage the development of the circular water economy that mitigates emissions while enhancing the resilience of communities and populations to climate change in the providing of water- and wastewater-related services. Instead, the majority of texts focus on the engineering and hard science aspects of the circular economy or on developments in particular facets of water resources management.

Developing the Circular Water Economy surveys the various technology policies used in leading locations to encourage the development of the circular water economy. The book will define the term 'technologies' according to the use by UNEP-DHI, which classifies technologies as hardware (physical infrastructure and technical equipment on the ground), software ('soft technologies' including approaches, processes, and methodologies including planning and decision support systems, models, knowledge transfer, and building capacity), and 'orgware' (organisation technologies for instance, organisation, ownership, and institutional arrangements).

Specifically, under each heading of reduce, reuse and recycle, and recover, *Developing the Circular Water Economy* surveys the various technology policies implemented by water utilities to develop the circular water economy that mitigates emissions and enhances resilience to climate change. Finally, readers are provided with an array of case studies of leading locations from around the world that have implemented policy solutions to encourage the emergence of the circular water economy.

Following this chapter, the book's chapter synopsis is as follows:

Chapter 2 discusses the impacts of climate change on water quantity and water quality.

Chapter 3 provides readers with an overview of the major, long-term non-climatic challenges to managing water in the linear water economy.

Chapter 4 introduces the concept of the circular water economy before discussing its contribution to mitigating emissions and enhancing resilience to climate change.

Chapters 5–7 provides readers with an understanding of the terms 'reduce', 'reuse' and 'recycle', and 'recover' in the context of the circular water economy.

Chapters 8–14 provides readers with case studies on the development of the circular water economy in a variety of locations around the world.

Chapter 15 provides readers with a series of best practices for other locations around the world developing the circular water economy.

Chapter 16 concludes from the case studies that developing the circular water economy requires innovative policies.

Notes

1. Edward B. Barbier, "The Role of Natural Resources in Economic Development," *Australian Economic Papers* 42, no. 2 (2003).

2. Neva Goodwin, Jonathan Harris, Julie A. Nelson, Brian Roach, and Mariano Torras, *Microeconomics in Context* (Armonk, NY: M. E. Sharpe, 2013).
3. World Economic Forum, "Towards the Circular Economy: Accelerating the Scale-Up Across Global Supply Chains," (2014), http://www3.weforum.org/docs/WEF_ENV_TowardsCircularEconomy_Report_2014.pdf.

References

Barbier, Edward B. "The Role of Natural Resources in Economic Development." *Australian Economic Papers* 42, no. 2 (2003): 253–72.
Brears, R. C. *Natural Resource Management and the Circular Economy*. Cham, Switzerland: Springer, 2018.
Goodwin, Neva, Jonathan Harris, Julie A. Nelson, Brian Roach, and Mariano Torras. *Microeconomics in Context*. Armonk, NY: M. E. Sharpe, 2013.
World Economic Forum. "Towards the Circular Economy: Accelerating the Scale-Up Across Global Supply Chains." (2014). http://www3.weforum.org/docs/WEF_ENV_TowardsCircularEconomy_Report_2014.pdf.

CHAPTER 2

Climatic Challenges to Water in the Linear Economy

Abstract In the linear economy, climatic challenges will impact both water quantity and water quality with decreased availability of surface and groundwater resources significantly intensifying while extreme weather events will pose threats to water quality. At the same time, providing of water- and wastewater-related services requires a lot of energy, and often results in greenhouse gases emitted.

Keywords Climate change · Greenhouse gas emissions · Carbon dioxide · Water quantity · Water quality

INTRODUCTION

Climate change will likely lead to higher temperatures and more extreme and less predictable weather conditions that will affect the availability and distribution of rainfall, snowmelt, surface water, and groundwater and further deteriorate water quality. This will impact the availability of good quality water of sufficient quantity for both humans and nature. This chapter will first discuss the impacts of climate change in a 1.5 °C and 2 °C above pre-industrial level world, before discussing the specific impacts of climate change on water quantity and water quality.

© The Author(s) 2020   5
R. C. Brears, *Developing the Circular Water Economy,*
Palgrave Studies in Climate Resilient Societies,
https://doi.org/10.1007/978-3-030-32575-6_2

A 1.5 °C or 2 °C World

Recognising that climate change represents an urgent and potentially irreversible threat to human societies and the planet, 195 countries at the Paris Climate Conference in December 2015 adopted the first-ever universal, legally binding global climate deal, known as the 'Paris Agreement'. The agreement sets out a global action plan to hold '... the increase in the global average temperature to well below 2 °C above pre-industrial levels' and to pursue '...efforts to limit the temperature increase to 1.5 °C above pre-industrial levels'. To limit warming to 1.5 °C with "no or limited overshoot", net global carbon emissions need to fall by around 45% from 2010 levels by 2030 and reach "net zero" by around 2050. This would require rapid and far-reaching transitions in energy, land, urban and infrastructure, and industrial systems. These system transitions are unprecedented in terms of scale and "would imply deep emissions reductions in all sectors, a wide portfolio of mitigation options and a significant upscaling of investments in these options".[1]

Human activities are estimated to have caused approximately 1.0 °C of global warming above pre-industrial levels. At current rates, human-caused warming is adding around 0.2 °C to global average temperatures every decade. If this trend continues, global warming is likely to reach 1.5 °C between 2030 and 2052. Climate-related risks for natural and human systems are higher for global warming of 1.5 °C than at present but lower than at 2 °C. On land, impacts on biodiversity and ecosystems, including species loss and extinction, are projected to be lower at 1.5 °C of global warming compared to 2 °C.[2]

Depending on future socio-economic conditions, limiting global warming to 1.5 °C compared to 2 °C may reduce the proportion of the world population exposed to climate change-induced increases in water stress by up to 50%. Populations living in river basins are projected to become newly exposed to chronic water scarcity even if global warming is constrained to less than 2 °C: out of the current population of around 1.3 billion exposed to water scarcity, about 3% (North America) to 9% (Europe) are expected to be prone to aggravated scarcity at 2 °C of global warming. Overall, exposure to the increase in water scarcity in 2050 will be globally reduced by 184–270 million people at about 1.5 °C of warming compared to the impacts of about 2 °C.[3]

Increases in the risks associated with runoff and in flood hazard can be expected at global warming of 1.5 °C, with an overall increase in the area affected by flooding hazard at 2 °C. Assuming constant population sizes, countries representing 73% of the world population will experience increasing flood risk, with a 100% increase at 1.5 °C and 170% increase at 2 °C compared to the impact simulated over the baseline period 1976–2005. The number of people exposed to increased flooding in 2050 under a 1.5 °C warming scenario could be reduced by 26–34 million compared to the number exposed to increased flooding associated with 2 °C of warming. In Europe, under a 2 °C global warming scenario with current socio-economic conditions and no adaptation measures in place, flood impacts could more than double, with around 525,000 people annually exposed to floods and €12.5 billion of expected annual losses.[4]

Under constant socio-economic conditions, the population exposed to drought at 2 °C of warming is projected to be larger than 1.5 °C: the global mean monthly number of people expected to be exposed to extreme drought at 1.5 °C in 2021–2040 is projected to be nearly 115 million, compared to just over 190 million at 2 °C in 2041–2060.[5] Climate change is likely to cause heatwaves to become more intense, extended, and frequent.[6] For example, in Florida climate change could result in summer heatwaves being three times more frequent, lasting six times longer than present, and getting hotter, rising roughly 4–6 °C.[7,8] Heatwaves lead to increased demand for water, for example, a heatwave in Tauranga, New Zealand led to a 30% increase in water consumption, from an average of 40 million litres per day to 52 million litres per day, resulting in water restrictions being put in place.[9]

Changes in climate and hydrology will have direct and cascading effects on water quality. Increases in high flow events can increase the delivery of sediment, nutrients, and microbial pathogens to streams, lakes, and estuaries, while decreases in flow volume during summer or droughts can impact aquatic life through exposure to high water temperatures and reduced dissolved oxygen. The risks of harmful algal blooms could increase due to an expanded seasonal window of warm water temperatures and the potential for episodic increases in nutrient loading from increases in winter precipitation and summer storms. In coastal areas, saltwater intrusion into coastal rivers and aquifers could be exacerbated by sea level rise and storm surges.[10,11]

SPECIFIC IMPACTS OF CLIMATE CHANGE
ON WATER QUANTITY AND WATER QUALITY

Freshwater-related risks of climate change increase significantly with increasing greenhouse gas concentrations: the proportion of the world's population experiencing water scarcity and major flooding events will increase with rising temperatures. Climate change is projected to decrease the availability of renewable surface water and groundwater resources significantly, intensifying competition for water resources among users. In presently dry regions droughts will likely increase, while precipitation is projected to increase at high latitudes. With extreme weather events (floods and droughts) climate change is projected to reduce the availability of good quality water and pose threats to drinking water quality, even with conventional treatment processes, due to interacting factors including increased temperature; increased sediment, nutrient and pollutant loadings from heavy rainfall; increased concentrations of pollutants during droughts; and disruption of treatment facilities during floods.[12,13] Specifically, climate change will impact water quantity and water quality in the following ways:

Increased Air Temperature

Changes in winter precipitation patterns will directly impact streamflow: an increase in average temperatures will cause more precipitation to fall as rain and earlier melt of the snow that does fall. Earlier snowmelt means lower flows later in the summer, placing stress on water supplies downstream.[14] Furthermore, increased evaporation rates from lakes, reservoirs, and aquifers will decrease the quantity of water available to users. At the same time, increased temperatures will increase demand for water.[15] In the western United States, snowmelt will start sooner in the season and at a slower rate because the warming occurs earlier when days are shorter with less sunlight. With 60 million people in the western United States dependent on snowmelt for their water supply, future declines in snowmelt-derived streamflow may place additional stress on over-allocated water supplies.[16]

Shifts in Timing of River Flows

Climate change is projected to decrease summer precipitation levels. At the same time, rising temperatures are projected to lead to more frequent and intense summer droughts. This will result in reduced river flows, which in turn will see a decline in water resources for various uses. In Europe, river flow droughts are projected to increase in frequency and severity in southern and south-eastern Europe, the United Kingdom, France, Benelux, and western parts of Germany over the coming decades.[17] Meanwhile, in southern Australia, an earlier study estimates that climate change will likely reduce average river flows by 10–20%.[18] As such, more frequent and intense droughts will increase the need for artificial water storage. Auxiliary infrastructure projects including dam augmentation by increasing dam height and aquifer storage and recovery through artificial recharge of deep and shallow aquifers are becoming more common to address issues of water availability while providing climate change adaptation.[19]

Higher Water Temperatures

Algal blooms usually develop during the spring when water temperature is higher and there is increased light. The growth is sustained during the warmer months of the year. Warmer water due to climate change may favour harmful algae as warmer water may prevent water from mixing, allowing algae to grow thicker and faster, and as algal blooms absorb sunlight it makes water even warmer, promoting more blooms. Increased algal blooms and natural organic material will lead to water needing additional or new treatment processes for drinking water.[20,21] The costs associated with treating algae and algal by-products in the Waco, Texas drinking water supply during 2002–2012 was estimated to be $70.4 million in addition to between $6.9 and 10.3 million in lost revenue from decreased water sales to neighbouring communities.[22]

Drier and Wetter Conditions Impacting Groundwater Quality

The effects of climate change on groundwater may include a long-term decline in groundwater storage, increased frequency, and severity, of groundwater droughts, increased frequency of groundwater-related floods, and mobilisation of pollutants due to seasonably high water tables.[23,24] For example, the San Joaquin aquifer in California has shrunk permanently by as much as 5.25% as a result of two drought periods: 2007–2009 and 2012–2015.[25] At the time, rural Californian communities reliant on wells are at risk from contaminants in the groundwater becoming more concentrated with less water available to dilute them, with many groundwater basins in the state contaminated from the overapplication of nitrogen fertiliser, industrial chemicals, or natural contaminants such as arsenic.[26] Meanwhile, in the Brahmaputra flood plain in Assam, India, researchers evaluating the heavy metal concentration status of groundwater in the area found concentrations of iron to have exceeded World Health Organization recommended levels of 0.3 mg/L in about 80% of samples while manganese values exceeded 0.4 mg/L in about 22.5% of samples and lead values exceeded the limit in 22.5% of samples.[27]

Increased Stormwater Run-Off

With rising global temperatures due to increased heat-trapping emissions, more water evaporates from the land and oceans. The warmer atmosphere can hold more water vapour. The result is the higher potential for heavy rainfall, which is the main cause of inland flooding. In Europe, floods have shifted considerably over the last 50 years with floods in the north-east of Europe, Sweden, Finland, and the Baltic States occurring one month earlier than in the 1960s and 1970s due to a warming climate.[28] Modern land-use practices, including increased development of floodplains, increased use of impermeable surfaces, and degradation of natural areas, means the landscape is less able to accommodate heavy rainfall, increasing the risks of floods and exacerbating their impacts.[29] Increased stormwater run-off will increase loads of pathogens, chemicals, nutrients, and suspended sediment. Climate change-induced precipitation changes alone could increase riverine total nitrogen loading by up to nearly 20% within the continental United States by the end of the century.[30] Increased suspended sediment can increase water treatment costs: In one study, a 1% decrease in turbidity leads to a $1123/year decrease in treatment costs.[31]

Sea-Level Rise

The quality of water supply in coastal and island regions is at risk from rising sea level and the occurrence of drought increasing the salinity of both surface water and groundwater through saltwater intrusion. As sea level rises, the 'salt front', which is the location of the freshwater-saltwater line, may progress further upstream. This encroachment may be further exacerbated by drought, reduced rainfall or changes in water use and demand. Saltwater intrusion can result in the need for water utilities to increase treatment, relocate water intake, or develop alternative sources of freshwater. Increased sea-level rises will increase the salinity of coastal aquifers particularly when groundwater recharge is predicted to decrease, comprising their usability. A concern of water managers is the potential adverse effect of sea level rise on the depth to the freshwater-saltwater interface near groundwater supply wells. Pumping of wells in coastal aquifers underlain by saltwater can lower the water table with respect to sea level, decreasing the depth to the freshwater-saltwater interface beneath the pumping well. This increases the potential for saltwater intrusion and potentially limits the amount of potable water available from the well.[32,33,34,35]

SUMMARY OF CLIMATE CHANGE IMPLICATIONS ON WATER RESOURCES

The summary of potential climate change and related extreme events and potential implications for water resources are listed in Table 2.1.[36]

Table 2.1 Summary of climate change implications on water resources

Changes in climate and extreme events	Potential implications for water supply, sewerage, drainage, and flooding	Increase/ decrease
Temperature increases	Annual demand for water	Increase
	Water quality impacts caused by turbidity from drier soils or vegetation dieback, pests, weeds, and increased pathogen growth	Increase
	Need to irrigate recreation areas, sporting, and parks/ gardens	Increase
	Snow, leading to less inflow to snowmelt-fed water storages in spring	Decrease
	Sewage quality changes and impacts on sewerage infrastructure e.g. corrosion and odour	Increase
More heatwaves	Threats to the health and safety of populations	Increase
	Power outages, which may affect water supply and sewerage services	Increase
	Water demand, potentially exceeding the capacity of the water grid	Increase
	Incidence of harmful algal blooms	Increase
Lower average rainfall	Streamflow and water quality in natural waterways	Decrease
	Quantity and quality of water available	Decrease
	Possible tensions between water supply for consumptive and environmental uses	Increase

(continued)

Table 2.1 (continued)

Changes in climate and extreme events	Potential implications for water supply, sewerage, drainage, and flooding	Increase/decrease
More droughts	Quantity and quality of water available for use	Decrease
	Recreational opportunities on water (e.g. fishing, swimming) and on land (e.g. sports grounds, private gardens)	Decrease
	Damage to water, sewerage, drainage, and flood management infrastructure due to dry soil which shifts and cracks, or tree roots seeking water sources	Increase
More intense rainfall	Quality of water for recreational, cultural, spiritual, and environmental uses when sewerage systems spill	Decrease
	Performance of drainage infrastructure	Decrease
	Runoff quality, especially from rainfall events after long dry periods	Decrease
	Flash flooding	Increase
	Storm damage to water sector assets	Increase

Contribution of the Linear Water Economy to Climate Change

Abstracting, treating, and conveying drinking water and treating wastewater requires a lot of energy. The energy used is mostly derived from traditional fossil fuel sources, such as coal, oil or natural gas. Energy production from these sources produces greenhouse gas emissions including carbon dioxide, methane, and nitrous oxide. Climate change is a direct challenge to water utilities as it can induce severe droughts and floods, affecting groundwater and surface water availability and quality, leading to increased abstraction-, treatment-, conveying-, and wastewater treatment-related emissions due to the reliance on more distant water, often of inferior quality.[37,38] Already, it is estimated that water utilities typically consume between 0.5 and 6% of regional energy produced and urban water management contributes up to 17% of regional greenhouse gas emissions.[39]

Notes

1. IPCC, "Special Report. Global Warming of 1.5 °C," (2018), https://www.ipcc.ch/sr15/.
2. Ibid.
3. Ibid.

4. J.-C. Ciscar Martinez, D. Ibarreta Ruiz, A. Soria Ramirez, A. Dosio, A. Toreti, A. Ceglar, D. Fumagalli, F. Dentener, R. Lecerf, A. Zucchini, L. Panarello, S. Niemeyer, I. Perez-Dominguez, T. Fellmann, A. Kitous, J. Després, A. Christodoulou, H. Demirel, L. Alfieri, F. Dottori, M. Vousdoukas, L. Mentaschi, E. Voukouvalas, C. Cammalleri, P. Barbosa, F. Micale, J. Vogt, J. I. Barredo Cano, G. Caudullo, A. Mauri, D. de Rigo, G. Libertà, T. Houston Durrant, T. Artés Vivancos, J. San-Miguel-Ayanz, S. Gosling, J. Zaherpour, A. de Roo, B. Bisselink, J. Bernhard, A. Bianchi, M. Rozsai, W. Szewczyk, I. Mongelli, L. Feyen, "Climate Impacts in Europe: Final Report of the JRC Peseta III Project," (2018), https://www.researchgate.net/publication/328956833_Climate_impacts_in_Europe_final_report_of_the_JRC_PESETA_III_project.
5. IPCC, "Special Report. Global Warming of 1.5 °C".
6. WMO, "2019 Starts with Extreme, High-Impact Weather," https://public.wmo.int/en/media/news/2019-starts-extreme-high-impact-weather.
7. Ajay Raghavendra et al., "Floridian Heatwaves and Extreme Precipitation: Future Climate Projections," *Climate Dynamics* 52, no. 1 (2019).
8. Phys.org, "Floridians Could Far Far More Frequent, Intense Heatwaves," https://phys.org/news/2018-05-floridians-frequent-intense-heatwaves.html.
9. New Zealand Herald, "Heatwave: Tauranga's Water Usage Up 30 Per cent to 52 Million Litres a Day," https://www.nzherald.co.nz/nz/news/article.cfm?c_id=1&objectid=12198492.
10. U. S. Global Change Research Program, "Impacts, Risks, and Adaptation in the United States: Fourth National Climate Assessment, Volume II," (2018), https://nca2018.globalchange.gov/.
11. Nigel W. Arnell et al., "The Implications of Climate Change for the Water Environment in England," *Progress in Physical Geography: Earth and Environment* 39, no. 1 (2015).
12. Ibid.
13. R. C. Brears, *Urban Water Security* (Chichester, UK; Hoboken, NJ: Wiley, 2016).
14. United States Department of Agriculture Forest Service, "Flows of the Future—How Will Climate Change Affect Streamflows in the Pacific Northwest?" (2016), https://www.fs.fed.us/pnw/sciencef/scifi187.pdf.
15. Brears, *Urban Water Security*.
16. Reno University of Nevada, "Slower Snowmelt Affects Downstream Water Availability in Western Mountains: Research Team Studying Areas with Significant Snowfall," Science, https://www.sciencedaily.com/releases/2016/08/160816182628.htm.

17. European Environment Agency, "River Flow Drought," https://www.eea.europa.eu/data-and-maps/indicators/river-flow-drought-1/assessment.
18. Francis Chiew, and Ian Prosser, "Water and Climate," In *Water: Science and Solutions for Australia*, edited by Ian Prosser. Clayton, Australia: CSIRO, 2011.
19. M. D. Perry and S. J. Praskievicz, "A New Era of Big Infrastructure? (Re) Developing Water Storage in the U.S. West in the Context of Climate Change and Environmental Regulation," *Water Alternatives* 10, no. 2 (2017).
20. Arnell et al., "The Implications of Climate Change for the Water Environment in England."
21. US EPA, "Climate Change and Harmful Algal Blooms," https://www.epa.gov/nutrientpollution/climate-change-and-harmful-algal-blooms.
22. Catherine R. Dunlap, Karen Seligman Sklenar, and Laura J. Blake, "A Costly Endeavor: Addressing Algae Problems in a Water Supply," *Journal—American Water Works Association* 107, no. 5 (2015).
23. UK Groundwater Forum, "Groundwater Resources and Climate Change," http://www.groundwateruk.org/groundwater_resources_climate_change.aspx.
24. IGRAC, "Groundwater & Climate Change," https://www.un-igrac.org/areas-expertise/groundwater-climate-change.
25. Chandrakanta Ojha, Susanna Werth, and Manoochehr Shirzaei, "Groundwater Loss and Aquifer System Compaction in San Joaquin Valley During 2012–2015 Drought, *Journal of Geophysical Research: Solid Earth* 124, no. 3 (2019).
26. Reuters, "Health Experts Warn of Water Contamination from California Drought," https://www.reuters.com/article/us-usa-california-drought/health-experts-warn-of-water-contamination-from-california-drought-idUSBREA1I06P20140219.
27. Nabanita Haloi, and H. P. Sarma, "Heavy Metal Contaminations in the Groundwater of Brahmaputra Flood Plain: An Assessment of Water Quality in Barpeta District, Assam (India)," *Environmental Monitoring and Assessment* 184, no. 10 (2012).
28. Günter Blöschl et al., "Changing Climate Shifts Timing of European Floods," *Science* 357, no. 6351 (2017).
29. Union of Concerned Scientists, "Climate Change, Extreme Precipitation and Flooding: The Latest Science (2018)," https://www.ucsusa.org/global-warming/global-warming-impacts/floods#.XBB5fWgzY2w.
30. E. Sinha, A. M. Michalak, and V. Balaji, "Eutrophication Will Increase During the 21st Century as a Result of Precipitation Changes," *Science* 357, no. 6349 (2017).

31. Matthew T. Heberling et al., "Comparing Drinking Water Treatment Costs to Source Water Protection Costs Using Time Series Analysis," *Water Resources Research* 51, no. 11 (2015).
32. Brears, *Urban Water Security.*
33. Richard G. Taylor et al., "Ground Water and Climate Change," *Nature Climate Change* 3 (2012).
34. US EPA, "Climate Impacts on Water Resources," https://archive.epa.gov/epa/climate-impacts/climate-impacts-water-resources.html.
35. "Climate Adaptation and Saltwater Intrusion," https://www.epa.gov/arc-x/climate-adaptation-and-saltwater-intrusion.
36. Land the State of Victoria Department of Environment, Water and Planning, "Pilot Water Sector Climate Change Adaptation Action Plan," (2018), https://www.water.vic.gov.au/__data/assets/pdf_file/0020/394220/WSAAP-Web-version-FINAL.pdf.
37. R. C. Brears, *The Green Economy and the Water-Energy-Food Nexus* (London: Palgrave Macmillan UK, 2017).
38. *Urban Water Security.*
39. WaCCLim, "The Roadmap to a Low-Carbon Urban Water Utility," (2018), http://wacclim.org/wp-content/uploads/2018/12/2018_WaCCliM_Roadmap_EN_SCREEN.pdf.

References

Arnell, Nigel W., Sarah J. Halliday, Richard W. Battarbee, Richard A. Skeffington, and Andrew J. Wade. "The Implications of Climate Change for the Water Environment in England." *Progress in Physical Geography: Earth and Environment* 39, no. 1 (2015): 93–120.
Blöschl, Günter, Julia Hall, Juraj Parajka, Rui A. P. Perdigão, Bruno Merz, Berit Arheimer, Giuseppe T. Aronica, et al. "Changing Climate Shifts Timing of European Floods." *Science* 357, no. 6351 (2017): 588.
Brears, R. C. *Urban Water Security.* Chichester, UK; Hoboken, NJ: Wiley, 2016.
———. *The Green Economy and the Water-Energy-Food Nexus.* London: Palgrave Macmillan UK, 2017.
Ciscar Martinez, J.-C., D. Ibarreta Ruiz, A. Soria Ramirez, A. Dosio, A. Toreti, A. Ceglar, D. Fumagalli, F. Dentener, R. Lecerf, A. Zucchini, L. Panarello, S. Niemeyer, I. Perez-Dominguez, T. Fellmann, A. Kitous, J. Després, A. Christodoulou, H. Demirel, L. Alfieri, F. Dottori, M. Vousdoukas, L. Mentaschi, E. Voukouvalas, C. Cammalleri, P. Barbosa, F. Micale, J. Vogt, J. I. Barredo Cano, G. Caudullo, A. Mauri, D. de Rigo, G. Libertà, T. Houston Durrant, T. Artés Vivancos, J. San-Miguel-Ayanz, S. Gosling, J. Zaherpour, A. de Roo, B. Bisselink, J. Bernhard, A. Bianchi, M. Rozsai, W. Szewczyk, I. Mongelli, L. Feyen. "Climate Impacts in Europe: Final Report

of the JRC Peseta III Project." (2018). https://www.researchgate.net/publication/328956833_Climate_impacts_in_Europe_final_report_of_the_JRC_PESETA_III_project.

Dunlap, Catherine R., Karen Seligman Sklenar, and Laura J. Blake. "A Costly Endeavor: Addressing Algae Problems in a Water Supply." *Journal—American Water Works Association* 107, no. 5 (2015, May 1): E255–E62.

European Environment Agency. "River Flow Drought." https://www.eea.europa.eu/data-and-maps/indicators/river-flow-drought-1/assessment.

Francis Chiew, and Ian Prosser. "Water and Climate." In *Water: Science and Solutions for Australia*, edited by Ian Prosser. Clayton, VI: CSIRO, 2011.

Haloi, Nabanita, and H. P. Sarma. "Heavy Metal Contaminations in the Groundwater of Brahmaputra Flood Plain: An Assessment of Water Quality in Barpeta District, Assam (India)." *Environmental Monitoring and Assessment* 184, no. 10 (2012, October 1): 6229–37.

Heberling, Matthew T., Christopher T. Nietch, Hale W. Thurston, Michael Elovitz, Kelly H. Birkenhauer, Srinivas Panguluri, Balaji Ramakrishnan, Eric Heiser, and Tim Neyer. "Comparing Drinking Water Treatment Costs to Source Water Protection Costs Using Time Series Analysis." *Water Resources Research* 51, no. 11 (2015, November 1): 8741–56.

IGRAC. "Groundwater & Climate Change." https://www.un-igrac.org/areas-expertise/groundwater-climate-change.

IPCC. "Special Report. Global Warming of 1.5 °C." (2018) https://www.ipcc.ch/sr15/.

New Zealand Herald. "Heatwave: Tauranga's Water Usage Up 30 Per cent to 52 Million Litres a Day." https://www.nzherald.co.nz/nz/news/article.cfm?c_id=1&objectid=12198492.

Ojha, Chandrakanta, Susanna Werth, and Manoochehr Shirzaei. "Groundwater Loss and Aquifer System Compaction in San Joaquin Valley During 2012–2015 Drought." *Journal of Geophysical Research: Solid Earth* 124, no. 3 (2019, March 1): 3127–43.

Perry, M. D., and S. J. Praskievicz, "A New Era of Big Infrastructure? (Re)Developing Water Storage in the U.S. West in the Context of Climate Change and Environmental Regulation." *Water Alternatives* 10, no. 2 (2017): 437–54.

Phys.org. "Floridians Could Far Far More Frequent, Intense Heatwaves." https://phys.org/news/2018-05-floridians-frequent-intense-heatwaves.html.

Raghavendra, Ajay, Aiguo Dai, Shawn M. Milrad, and Shealynn R. Cloutier-Bisbee. "Floridian Heatwaves and Extreme Precipitation: Future Climate Projections." *Climate Dynamics* 52, no. 1 (2019, January 1): 495–508.

Reuters. "Health Experts Warn of Water Contamination from California Drought." https://www.reuters.com/article/us-usa-california-drought/health-experts-warn-of-water-contamination-from-california-drought-idUS-BREA1I06P20140219.

Sinha, E., A. M. Michalak, and V. Balaji. "Eutrophication Will Increase During the 21st Century as a Result of Precipitation Changes." *Science* 357, no. 6349 (2017): 405.

Taylor, Richard G., Bridget Scanlon, Petra Döll, Matt Rodell, Rens van Beek, Yoshihide Wada, Laurent Longuevergne, et al. "Ground Water and Climate Change." *Nature Climate Change* 3 (2012, November 25): 322.

The State of Victoria Department of Environment, Land, Water and Planning. "Pilot Water Sector Climate Change Adaptation Action Plan." (2018). https://www.water.vic.gov.au/__data/assets/pdf_file/0020/394220/WSAAP-Web-version-FINAL.pdf.

UK Groundwater Forum. "Groundwater Resources and Climate Change." http://www.groundwateruk.org/groundwater_resources_climate_change.aspx.

Union of Concerned Scientists. "Climate Change, Extreme Precipitation and Flooding: The Latest Science (2018)." https://www.ucsusa.org/global-warming/global-warming-impacts/floods#.XBB5fWgzY2w.

United States Department of Agriculture Forest Service. "Flows of the Future—How Will Climate Change Affect Streamflows in the Pacific Northwest?" (2016). https://www.fs.fed.us/pnw/sciencef/scifi187.pdf.

University of Nevada, Reno. "Slower Snowmelt Affects Downstream Water Availability in Western Mountains: Research Team Studying Areas with Significant Snowfall." Science. https://www.sciencedaily.com/releases/2016/08/160816182628.htm.

US EPA. "Climate Adaptation and Saltwater Intrusion." https://www.epa.gov/arc-x/climate-adaptation-and-saltwater-intrusion.

———. "Climate Change and Harmful Algal Blooms." https://www.epa.gov/nutrientpollution/climate-change-and-harmful-algal-blooms.

———. "Climate Impacts on Water Resources." https://archive.epa.gov/epa/climate-impacts/climate-impacts-water-resources.html.

U.S. Global Change Research Program. "Impacts, Risks, and Adaptation in the United States: Fourth National Climate Assessment, Volume II." (2018). https://nca2018.globalchange.gov/.

WaCCLim. "The Roadmap to a Low-Carbon Urban Water Utility." (2018). http://wacclim.org/wp-content/uploads/2018/12/2018_WaCCliM_Roadmap_EN_SCREEN.pdf.

WMO. "2019 Starts with Extreme, High-Impact Weather." https://public.wmo.int/en/media/news/2019-starts-extreme-high-impact-weather.

CHAPTER 3

Non-climatic Challenges
to Water in the Linear Economy

Abstract There are a variety of major, long-term non-climatic challenges to managing water in the linear economy including population growth and rapid urbanisation, economic growth and rising income levels, more stringent environmental legislation, the water-energy and water-food nexus, ageing infrastructure, and changing customer expectations.

Keywords Resource scarcity · Water scarcity · Environmental degradation · Water-energy · Water-food

INTRODUCTION

While economic growth in the linear economy has produced many benefits including raising living standards and improving quality of life around the world, it has also resulted in water quantity and water quality impacts that are detrimental to human health as well as the health of ecosystems. Specific long-term non-climatic challenges to water quantity and water quality in the linear economy include population growth and rapid urbanisation, economic growth and rising income levels, more stringent environmental legislation, the water-energy and water-food nexus, and ageing infrastructure. In addition, there are rising customer expectations for the delivery of water-related services.

© The Author(s) 2020
R. C. Brears, *Developing the Circular Water Economy*,
Palgrave Studies in Climate Resilient Societies,
https://doi.org/10.1007/978-3-030-32575-6_3

POPULATION GROWTH

In 2017, the world's population reached 7.6 billion, with the world having added an extra one billion people over the past 12 years. Over the next 13 years, it is projected that more than one billion people will be added, reaching 8.6 billion in 2030, and increasing further to 9.8 billion in 2050 and 11.2 billion by 2100.[1] Globally, demand for water has not been linear, with demand for water estimated to have been double the rate of population growth over the last century.[2] Already, almost one-fifth of the world's population, around 1.2 billion people, live in areas where water is physically scarce. Population growth will impact ecosystems around the world with direct drivers of ecosystem degradation including excessive water withdrawal, eutrophication, and pollution.[3,4]

RAPID URBANISATION

Today, 55% of the world's population lives in cities and by 2050 this will rise to 68%. Urbanisation, combined with the overall growth of the world's population, could add another 2.5 billion people to urban areas by 2050. Currently, there are 33 cities with more than 10 million inhabitants. By 2030, the world is projected to have 43 mega-cities with more than 10 million inhabitants.[5]

Cities impact the hydrological cycle by extracting significant amounts of water from surface and groundwater sources and extending impervious surfaces, which prevents recharge of groundwater, exacerbates flood risks, and pollutes water bodies through discharge of untreated wastewater. Regarding water scarcity, in Dhaka city, Bangladesh, the city's population is increasing by 5% per annum while water demand is increasing at 4% annually. With 87% of the city's water supply based on groundwater resources, the rise in water consumption has seen the groundwater level fall from 28 metres in 1997 to 70 metres in 2012.[6] Meanwhile, on average only 20% of produced wastewater globally receives proper treatment, with treatment level dependent on the income level of the country: treatment capacity is 70% of the generated wastewater in high-income countries compared to only 8% in low-income countries.[7]

Peri-Urban Water Competition

Much of the water consumed by cities generally comes from outside the city limits. In one study, it is estimated that, globally, large cities obtain around 78% of their water from surface sources and that, cumulatively, large cities moved 504 billion litres a day a distance of about 27,000 kilometres, with the upstream contributing area of urban water sources calculated at 41% of the global land surface.[8] However, water for cities will become scarcer as the century progresses: reduced freshwater availability and competition from other uses including energy and agriculture could reduce water availability in cities by as much as two thirds by 2050, compared to 2015 levels.[9]

Rapid Economic Growth and Rising Income Levels

The world's economy is projected to double in size by 2037 and nearly triple by 2050.[10] While this growth will result in a less than proportional increase in water demand, the global economy will still require 55% more water.[11] Global water demand for manufacturing is projected to increase by 400% from 2000 to 2050.[12] Meanwhile the proportion of water required by industry will increase: approximately 20% of the world's freshwater resources are used by industry with the percentage of a country's industrial sector's water demand proportional to the average income level, ranging from around 5% of water withdrawals in low-income countries to over 40% in some high-income countries.[13] However, economic growth is dependent on the hydrological cycle with a 1% increase in drought area leading to a 2.8% reduction in economic growth, while a 1% increase in area impacted by floods results in a 1.8% decrease in economic growth.[14]

Household water demand is projected to increase by 130% due to higher incomes and living standards.[15,16] Almost three billion people, more than 40% of the world's current population, will join the middle classes by 2050. As a result, emerging markets will comprise almost two-thirds of global consumption by 2050, compared to one-third today, impacting global water consumption patterns.[17] One of the main expected changes in rising incomes in emerging economies is a shift in diet from predominantly starch-based to water-intensive meat and dairy products.[18,19]

ENVIRONMENTAL LEGISLATION

In order to address growing environmental concerns, governments have introduced and strengthened policies that implicitly or explicitly increase the price of using the environment as a factor of production. Potable and wastewater service providers are being asked to meet ever-increasing levels of reliability, quality, and resilience. For instance, in the United Kingdom, the Priority Substances Directive supplements the European Union Water Framework Directive by establishing environmental quality standards for 'priority substances' and 'priority hazardous substances' with targets already in place for some substances, with future targets agreed for additional substances. This list of substances is updated periodically and there are a number of additional substances on the watch list for potential future inclusion.[20] The costs of meeting these service levels are likely to place pressure on affordability as these costs are all expected to be recovered through user charges. These costs may make the services unaffordable for some communities, particularly low-income communities as well as ones where local populations are ageing and/or in decline.[21,22] For example, councils across New Zealand face a potential bill of around NZD$500 million to meet stringent new drinking water quality standards, with the Canterbury region alone requiring NZD$200 million to ensure its water treatment plants are adequate; however, these costs may be unaffordable for many communities, particularly smaller communities.[23]

THE WATER-ENERGY NEXUS

Water requires energy for a range of water processes, including water distribution, wastewater treatment, and desalination. At the same time, energy requires water in all phases of energy production, including fossil fuels, biofuels, and power plants. Currently, the amount of energy used in the water sector is equivalent to the entire energy demand of Australia. In 2014, around 4% of global electricity consumption was used to extract, distribute, and treat water and wastewater, along with 50 million tonnes of oil equivalent of thermal energy, mostly diesel used for irrigation pumps and gas in desalination plants. Between now and 2040, it is projected that the amount of energy used in the water sector will double, with the largest increase coming from desalination, followed by large-scale water transfer and increasing demand for wastewater treatment as well as higher levels of treatment.[24] For most municipal governments, water and wastewater plants are typically the largest energy consumers, often accounting for 30–40% of total energy consumed. Energy prices can have a substantial impact on utility costs (as much as 40% of operating costs for drinking water systems can be for energy[25]) and fluctuations can make accurate forecasting difficult. For example, over the period 2006–2016, residential electricity rates in Ontario, Canada increased between 71 and 149% while Alberta, Canada, recalculates its electricity rate monthly and so electricity costs for water utilities can fluctuate significantly. In addition, future energy costs for water utilities are expected to rise due to the energy requirements of meeting more stringent water treatment standards as well as treating water of lower quality due to human activities including urbanisation and climate change.[26] Meanwhile, the energy sector is responsible for 10% of global water withdrawals. It is estimated that by 2040, water withdrawals for the sector could increase by less than 2% to 400 billion cubic metres (bcm) while the amount of water consumed (withdrawn but not returned to source) could increase by over 60% to over 75 bcm.[27]

The Water-Food Nexus

Agriculture accounts for 70% of all water withdrawn.[28] Annual global agricultural water consumption includes crop water consumption for food, fibre, and seed production plus evaporation losses from the soil and open water associated with agriculture, for example, rice fields, irrigation canals, and reservoirs. By 2050, the world will require at the minimum 60% more food produce to maintain current consumption patterns. This will result in the volume of global water withdrawn for irrigation increasing from 2.6 billion cubic kilometres in 2005–2007 to 2.9 billion cubic kilometres in 2050.[29]

The increase in agricultural production will impact the quality of water resources due to non-point source pollution. Key problems include sediment runoff that causes siltation problems; nutrient runoff with nitrogen and phosphorous being key pollutants found in agricultural runoff having been applied to farmland in several ways including as a fertiliser, animal manure, and municipal wastewater; microbial runoff from livestock or use of excreta as fertiliser; and chemical runoff from pesticides, herbicides, and other agrichemicals contaminating surface water and groundwater. The Gulf of Mexico dead zone, caused by nutrient enrichment from the Mississippi River, particularly nitrogen and phosphorous from the major farming states, is an area of hypoxic (less than two parts per million dissolved oxygen) waters at the mouth of the Mississippi River. The dead zone varies in size and can cover up to 6000–7000 square miles. In 2017, the largest ever dead zone was recorded with it becoming the size of New Jersey.[30]

Around 220 million tonnes of phosphate rock is mined worldwide every year, with phosphate rock geographically concentrated in five countries (Morocco, China, Algeria, Syria, and Jordan), which combined control 85–90% of the world's remaining reserves. This makes other regions of the word vulnerable to phosphate scarcity, volatile pricing, monopolisation, and geopolitical tensions. At the same time, 'peak phosphorous' describes the time at which the maximum possible global phosphorous rock production rate is reached, the quality of the remaining reserves decreases (more contaminants, less phosphorous), and they become harder to access (for example, moving from land-based deposits to ocean-based ones). This would make it uneconomical to mine and process, resulting in a decline in supply with a corresponding rise in price. The most likely time it is believed peak phosphorous will be reached is 2100.[31]

Ageing Infrastructure

In many locations around the world, significant new investment and increased efficiencies are needed as water filtration plants, pipes, and pumps age past their useful life. In addition to ageing infrastructure leading to leakage, broken or blocked wastewater pipes may discharge untreated sewage into local waterways threatening human health. In the United States, nearly six billion gallons of treated drinking water is lost due to leaking pipes, with an estimated 240,000 water main breaks occurring each year. It is estimated that leaky, ageing pipes are wasting 14–18% of each day's treated water, which is enough to support 15 million households.[32] In South Africa, it is estimated that on average 37% of the country's water supply is lost before it reaches users due to leaks with infrastructure in the rural areas most at risk of failure.[33] Regarding the United States, the American Water Works Association estimates that restoring existing water systems as they reach the end of their useful lives and expanding them to serve a growing population will cost at least $1 trillion over the next 25 years: this is just to maintain current levels of service. Similarly, in Europe, there are many segments of the European

Union's seven million kilometres of pipes that have been in operation for over 100 years with investment in water infrastructure not keeping up with growing populations, urbanisation or climate change with the sector needing to double its annual investment of €45 billion to modernise its infrastructure, protect health and the environment, and reduce costs.[34] Overall, delaying investment in water infrastructure could mean either more frequent rates of pipe breakages and deteriorating water service or sub-optimal use of utility funds, for instance paying more to repair broken pipes than the long-term cost of replacing them.[35]

CHANGING CUSTOMER PROFILES AND EXPECTATIONS

Water utilities are under increasing pressure to show value for rates paid and to enhance customer engagement and participation in various programmes. This has resulted in end users of water services transitioning from being captive consumers of a uniform product delivered under fixed circumstances to end users that demand they be able to choose different products and services, for example, purchasing rain-water harvesting systems and having access to real-time information on customer metering data to determine how much water is being used and whether there is a leak.[36] This effectively turns the consumer into a co-constructor of new water infrastructure, helping to support water innovations while at the same time demanding these systems be delivered and subsidised by the water utility or municipal agencies. Furthermore, water users are demanding that global water-using practices become more sustainable, which in turn provides support to water conservation initiatives developed by their local water utilities.[37]

NOTES

1. Department of Economic and Social Affairs United Nations, Population Division, "World Population Prospects: The 2017 Revision, Key Findings and Advance Tables," (2017), https://esa.un.org/unpd/wpp/Publications/Files/WPP2017_KeyFindings.pdf.
2. OECD, "Pricing Water Resources and Water and Sanitation Services," (2010), https://www.oecd-ilibrary.org/environment/pricing-water-resources-and-water-and-sanitation-services_9789264083608-en.
3. R. C. Brears, *Urban Water Security* (Chichester, UK; Hoboken, NJ: Wiley, 2016).

4. *Blue and Green Cities: The Role of Blue-Green Infrastructure in Managing Urban Water Resources* (Palgrave Macmillan UK, 2018).
5. Population Division UN Department of Economic and Social Affairs, "2018 Revision of World Urbanization Prospects," (2018), https://www.un.org/development/desa/publications/2018-revision-of-world-urbanization-prospects.html.
6. Md Arfanuzzaman, and A. Atiq Rahman, "Sustainable Water Demand Management in the Face of Rapid Urbanization and Ground Water Depletion for Social-Ecological Resilience Building," *Global Ecology and Conservation* 10 (2017).
7. UN-Water, "Wastewater Management: A UN-Water Analytical Brief," (2015), http://www.unwater.org/publications/wastewater-management-un-water-analytical-brief/.
8. Robert I. McDonald et al., "Water on an Urban Planet: Urbanization and the Reach of Urban Water Infrastructure," *Global Environmental Change* 27 (2014).
9. World Bank, "High and Dry: Climate Change, Water, and the Economy," (2016), https://openknowledge.worldbank.org/handle/10986/23665?utm_source=Global+Waters+%2B+Water+Currents&utm_campaign=9905bbdc1e-Water_Currents_Water+Utiliti_12_dec_2018&utm_medium=email&utm_term=0_fae9f9ae2b-9905bbdc1e-25803553.
10. PwC, "The World in 2050: Will the Shift in Global Economic Power Continue?" (2015), http://www.pwc.com/gx/en/issues/the-economy/assets/world-in-2050-february-2015.pdf.
11. UN-Water, "Partnerships for Improving Water and Energy Access, Efficiency and Sustainability," (2014), http://www.un.org/waterforlifedecade/water_and_energy_2014/pdf/water_and_energy_2014_final_report.pdf.
12. OECD, "OECD Environmental Outlook to 2050: The Consequences of Inaction," (2012), http://www.oecd.org/env/indicators-modelling-outlooks/oecd-environmental-outlook-1999155x.htm.
13. UNESCO, "Managing Water Under Uncertainty and Risk," (2012), http://www.unesco.org/new/fileadmin/MULTIMEDIA/HQ/SC/pdf/WWDR4%20Volume%201-Managing%20Water%20under%20Uncertainty%20and%20Risk.pdf.
14. UN-Water, "A Post-2015 Global Goal for Water: Synthesis of Key Findings and Recommendations from UN-Water," (2014), http://www.un.org/waterforlifedecade/pdf/27_01_2014_un-water_paper_on_a_post2015_global_goal_for_water.pdf.
15. Sharon L. Harlan et al., "Household Water Consumption in an Arid City: Affluence, Affordance, and Attitudes," *Society & Natural Resources* 22, no. 8 (2009).

16. UN-Water, "Partnerships for Improving Water and Energy Access, Efficiency and Sustainability".
17. HSBC, "Consumer in 2050—The Rise of the Em Middle Class," (2012), https://www.hsbc.com.vn/1/PA_ES_Content_Mgmt/content/vietnam/abouthsbc/newsroom/attached_files/HSBC_report_Consumer_in_2050_EN.pdf.
18. John Kearney, "Food Consumption Trends and Drivers," *Philosophical Transactions of the Royal Society of London. Series B, Biological Sciences* 365, no. 1554 (2010).
19. UNESCO, "Managing Water Under Uncertainty and Risk".
20. Ofwat, "Towards Water 2020—Meeting the Challenges for Water and Wastewater Services in England and Wales," (2015), https://www.ofwat.gov.uk/wp-content/uploads/2015/10/pap_tec201507challenges.pdf.
21. Local Government New Zealand, "Improving New Zealand's Water, Wastewater and Stormwater Sector," (2015), http://www.lgnz.co.nz/assets/Uploads/29617-three-Waters-Position-Paper.pdf.
22. Yorkshire Water, "Water 2020 Issues Paper—Summary," (2015), https://www.yorkshirewater.com/sites/default/files/Issues%20Paper%20%28Future%20Challenges%29%20v0.pdf.
23. Stuff, "Councils Facing $500 m Bill to Meet Stringent New Recommendations on Drinking Water Quality," https://www.stuff.co.nz/environment/103453009/councils-facing-500m-bill-to-meet-stringent-new-recommendations-on-drinking-water-quality.
24. IEA, "World Energy Outlook 2016: Water-Energy Nexus Excerpt," (2016), https://www.iea.org/weo/weospecialreports/.
25. US EPA, "Energy Efficiency for Water Utilities," https://www.epa.gov/sustainable-water-infrastructure/energy-efficiency-water-utilities.
26. Canadian Water Network, "Balancing the Books: Financial Sustainability for Canadian Water Systems," (2018), http://cwn-rce.ca/report/balancing-the-books-financial-sustainability-for-canadian-water-systems/.
27. IEA, "World Energy Outlook 2016: Water-Energy Nexus Excerpt".
28. FAO, "The Future of Food and Agriculture-Trends and Challenges," (2017), http://www.fao.org/3/a-i6583e.pdf.
29. "Towards a Water and Food Secure Future: Critical Perspectives for Policy-Makers," (2015), http://www.fao.org/3/a-i4560e.pdf.
30. Carleton College, "What Is the Gulf of Mexico Dead Zone?" https://serc.carleton.edu/microbelife/topics/deadzone/index.html.
31. EPA Ireland, "Identification and Evaluation of Phosphorus Recovery Technologies in an Irish Context," (2016), http://www.epa.ie/researchandeducation/research/researchpublications/researchreports/research189.html.

32. ASCE, "2017 Infrastructure Report Card," (2018), https://www.infra-structurereportcard.org/.
33. Fin24, "Over a Third of SA Water Supply Lost Through Poor Infrastructure," https://www.fin24.com/Economy/South-Africa/over-a-third-of-sa-water-supply-lost-through-poor-infrastructure-20180316.
34. Euractiv, "Time to Invest in Europe's Water Infrastructure," https://www.euractiv.com/section/energy-environment/opinion/time-to-invest-in-europes-water-infrastructure/.
35. AWWA, "Buried No Longer: Confronting America's Water Infrastructure Challenge," (2017), https://www.awwa.org/Portals/0/files/legreg/documents/BuriedNoLonger.pdf.
36. NACWA, "Envisioning the Digital Utility of the Future," (2017), http://www.nacwa.org/docs/default-source/conferences-events/2017-summer/17ulc-digital-utility-r6.pdf?sfvrsn=2.
37. D. L. T. Hegger et al., "Consumer-Inclusive Innovation Strategies for the Dutch Water Supply Sector: Opportunities for More Sustainable Products and Services," *NJAS—Wageningen Journal of Life Sciences* 58, no. 1 (2011).

References

Arfanuzzaman, Md, and A. Atiq Rahman. "Sustainable Water Demand Management in the Face of Rapid Urbanization and Ground Water Depletion for Social-Ecological Resilience Building." *Global Ecology and Conservation* 10 (2017, April 1): 9–22.

ASCE. "2017 Infrastructure Report Card." (2018). https://www.infrastructurereportcard.org/.

AWWA. "Buried No Longer: Confronting America's Water Infrastructure Challenge." (2017). https://www.awwa.org/Portals/0/files/legreg/documents/BuriedNoLonger.pdf.

Brears, R. C. *Blue and Green Cities: The Role of Blue-Green Infrastructure in Managing Urban Water Resources.* London: Palgrave Macmillan UK, 2018.

———. *Urban Water Security.* Chichester, UK; Hoboken, NJ: Wiley, 2016.

Canadian Water Network. "Balancing the Books: Financial Sustainability for Canadian Water Systems." (2018). http://cwn-rce.ca/report/balancing-the-books-financial-sustainability-for-canadian-water-systems/.

Carleton College. "What Is the Gulf of Mexico Dead Zone?" https://serc.carleton.edu/microbelife/topics/deadzone/index.html.

EPA Ireland. "Identification and Evaluation of Phosphorus Recovery Technologies in an Irish Context." (2016). http://www.epa.ie/researchandeducation/research/researchpublications/researchreports/research189.html.

Euractiv. "Time to Invest in Europe's Water Infrastructure." https:// www.euractiv.com/section/energy-environment/opinion/ time-to-invest-in-europes-water-infrastructure/.

FAO. "The Future of Food and Agriculture-Trends and Challenges." (2017). http://www.fao.org/3/a-i6583e.pdf.

———. "Towards a Water and Food Secure Future: Critical Perspectives for Policy-Makers." (2015). http://www.fao.org/3/a-i4560e.pdf.

Fin24. "Over a Third of SA Water Supply Lost Through Poor Infrastructure." https://www.fin24.com/Economy/South-Africa/over-a-third-of-sa-water-supply-lost-through-poor-infrastructure-20180316.

Harlan, Sharon L., Scott T. Yabiku, Larissa Larsen, and Anthony J. Brazel. "Household Water Consumption in an Arid City: Affluence, Affordance, and Attitudes." *Society & Natural Resources* 22, no. 8 (2009, August 11): 691–709.

Hegger, D. L. T., G. Spaargaren, B. J. M. van Vliet, and J. Frijns. "Consumer-Inclusive Innovation Strategies for the Dutch Water Supply Sector: Opportunities for More Sustainable Products and Services." *NJAS—Wageningen Journal of Life Sciences* 58, no. 1 (2011, June 1): 49–56.

HSBC. "Consumer in 2050—The Rise of the Em Middle Class." (2012). https://www.hsbc.com.vn/1/PA_ES_Content_Mgmt/content/vietnam/ abouthsbc/newsroom/attached_files/HSBC_report_Consumer_in_ 2050_EN.pdf.

IEA. "World Energy Outlook 2016: Water-Energy Nexus Excerpt." (2016). https://www.iea.org/weo/weospecialreports/.

Kearney, John. "Food Consumption Trends and Drivers." *Philosophical Transactions of the Royal Society of London. Series B, Biological Sciences* 365, no. 1554 (2010): 2793–807.

Local Government New Zealand. "Improving New Zealand's Water, Wastewater and Stormwater Sector." (2015). http://www.lgnz.co.nz/assets/Uploads/ 29617-three-Waters-Position-Paper.pdf.

McDonald, Robert I., Katherine Weber, Julie Padowski, Martina Flörke, Christof Schneider, Pamela A. Green, Thomas Gleeson, et al. "Water on an Urban Planet: Urbanization and the Reach of Urban Water Infrastructure." *Global Environmental Change* 27 (2014): 96–105.

NACWA. "Envisioning the Digital Utility of the Future." (2017). http:// www.nacwa.org/docs/default-source/conferences-events/2017-sum-mer/17ulc-digital-utility-r6.pdf?sfvrsn=2.

OECD. "OECD Environmental Outlook to 2050: The Consequences of Inaction." (2012). http://www.oecd.org/env/indicators-modelling-out-looks/oecd-environmental-outlook-1999155x.htm.

———. "Pricing Water Resources and Water and Sanitation Services." (2010). https://www.oecd-ilibrary.org/environment/pricing-water-resources-and-water-and-sanitation-services_9789264083608-en.

Ofwat. "Towards Water 2020—Meeting the Challenges for Water and Wastewater Services in England and Wales." (2015). https://www.ofwat.gov.uk/wp-content/uploads/2015/10/pap_tec201507challenges.pdf.

PwC. "The World in 2050: Will the Shift in Global Economic Power Continue?" (2015). http://www.pwc.com/gx/en/issues/the-economy/assets/world-in-2050-february-2015.pdf.

Stuff. "Councils Facing $500 m Bill to Meet Stringent New Recommendations on Drinking Water Quality." https://www.stuff.co.nz/environment/103453009/councils-facing-500m-bill-to-meet-stringent-new-recommendations-on-drinking-water-quality.

UN Department of Economic and Social Affairs, Population Division. "2018 Revision of World Urbanization Prospects." (2018). https://www.un.org/development/desa/publications/2018-revision-of-world-urbanization-prospects.html.

UNESCO. "Managing Water Under Uncertainty and risk." (2012). http://www.unesco.org/new/fileadmin/MULTIMEDIA/HQ/SC/pdf/WWDR4%20Volume%201-Managing%20Water%20under%20Uncertainty%20and%20Risk.pdf.

United Nations, Department of Economic and Social Affairs, Population Division. "World Population Prospects: The 2017 Revision, Key Findings and Advance Tables." (2017). https://esa.un.org/unpd/wpp/Publications/Files/WPP2017_KeyFindings.pdf.

UN-Water. "Partnerships for Improving Water and Energy Access, Efficiency and Sustainability." (2014). http://www.un.org/waterforlifedecade/water_and_energy_2014/pdf/water_and_energy_2014_final_report.pdf.

———. "A Post-2015 Global Goal for Water: Synthesis of Key Findings and Recommendations from UN-Water." (2014). http://www.un.org/waterforlifedecade/pdf/27_01_2014_un-water_paper_on_a_post2015_global_goal_for_water.pdf.

———. "Wastewater Management: A UN-Water Analytical Brief." (2015). http://www.unwater.org/publications/wastewater-management-un-water-analytical-brief/.

US EPA. "Energy Efficiency for Water Utilities." https://www.epa.gov/sustainable-water-infrastructure/energy-efficiency-water-utilities.

World Bank. "High and Dry: Climate Change, Water, and the Economy." (2016). https://openknowledge.worldbank.org/handle/10986/23665?utm_source=Global+Waters+%2B+Water+Currents&utm_campaign=9905bbdc1e-Water_Currents_Water+Utiliti_12_dec_2018&utm_medium=email&utm_term=0_fae9f9ae2b-9905bbdc1e-25803553.

Yorkshire Water. "Water 2020 Issues Paper—Summary." (2015). https://www.yorkshirewater.com/sites/default/files/Issues%20Paper%20%28Future%20Challenges%29%20v0.pdf.

CHAPTER 4

The Circular Water Economy

Abstract In the circular economy, the aim is to keep resources in use for as long as possible, extract value from them while in use, and recover and regenerate products and materials at the end of each service life. The circular water economy optimises water resources and extracts valuable resources from water and wastewater, all the while mitigating emissions and enhancing resilience to climate change.

Keywords Circular economy · Circular water economy · Climate change · Mitigation · Resilience · Resource efficiency

INTRODUCTION

In the circular economy, the aim is to keep resources in use for as long as possible, extract value from them while in use, and recover and regenerate products and materials at the end of each service life.[1] As such, the circular economy focuses on recycling, limiting, and reusing the physical inputs of the economy and using waste as a resource, leading to reduced primary resource consumption.[2] This is commonly referred to as the 3R (reduce, reuse, and recycle) approach. A key aspect of this approach is that materials, which have accumulated in the circular economy, constitute important man-made stocks that can be exploited through recycling to gain secondary raw materials and reused and remanufactured to keep

© The Author(s) 2020

R. C. Brears, *Developing the Circular Water Economy*,
Palgrave Studies in Climate Resilient Societies,
https://doi.org/10.1007/978-3-030-32575-6_4

products in the commercial life cycle.[3,4] Based on this, the chapter will first discuss the linear economy before discussing the concept of the circular water economy and its contribution to mitigating emissions and enhancing resilience to climate change.

THE LINEAR ECONOMY

In our current economic model, manufactured capital, human capital, and natural capital all contribute to human welfare by supporting the production of goods and services in the economic process, where natural capital—the world's stock of natural resources (provided by nature before their extraction or processing by humans)—is typically used for material and energy inputs into production and acts as a 'sink' for waste from the economic process.[5] This economic model can be best described as 'linear' which typically involves economic actors[6]—who are people or organisations engaged in any of the four economic activities of production, distribution, consumption, and resource maintenance—harvesting and extracting natural resources, using them to manufacture a product, and selling a product to other economic actors, who then discard it when it no longer serves its purpose.[7] Thus, natural resources in the linear economic model:

- *Become inputs*: Material resources used in the economy come from raw materials that are extracted from domestic natural resource stocks or extracted from natural resource stocks abroad and imported in the form of raw materials, semi-finished materials, or materials embedded in manufactured goods. Material resources are extracted with the usable parts of the resources entering the economy as material inputs where they become priced goods that are traded, processed, and used. Other parts remain unused in the environment and are called unused materials or unused extraction.
- *Become outputs*: After use in production and consumption activities, materials leave the economy as an output either to the environment in the form of residuals (pollution, waste) or in the form of raw materials, semi-finished materials, and materials embedded in manufactured goods.
- *Accumulate in man-made stocks*: Some materials accumulate in the Economy where they are stored in the form of buildings, transport infrastructure, or durable and semi-durable goods such as cars,

industrial machinery, and household appliances. These materials are eventually released in the form of demolition waste, end-of-life vehicles, e-waste, bulky household waste, and so on, which if not recovered flow back to the environment.

- *Create indirect flows.* When materials or goods are imported for use in an economy, their upstream production is associated with unused materials that remain abroad including raw materials needed to produce the goods and the generation of residuals. These indirect flows of materials consider the life cycle dimension of the production chain but are not physically imported. As such, the environmental consequences occur in countries from which the imports originate.[8]

THE CIRCULAR ECONOMY DECOUPLING ECONOMIC GROWTH FROM RESOURCE USE

The circular economy, in contrast to the linear 'take-make-consume-dispose' economy, aims to decouple economic growth from resource use and associated environmental impacts. The notion of decoupling is that economic output shall continue to increase at the same time as rates of increasing resource use and environmental impact are slowed, and in time brought into decline.[9] Specifically, decoupling is said to be absolute when the environmental variable is stable or decreasing while the economic variable is growing. Decoupling is considered relative when the environmental variable is increasing but at a lower rate than the economic variable. In the context of natural resources management, two modes of decoupling can be distinguished:

- Resource decoupling or 'dematerialisation' involves reducing the rate at which natural resources are used per unit of economic output.
- Impact decoupling seeks to increase economic activity while decreasing negative environmental impacts from pressures such as pollution, carbon emissions, or destruction of biodiversity.[10]

THE CIRCULAR WATER ECONOMY

In the development of the circular water economy, the aim is to design out externalities and keep resources in use, all the while regenerating natural capital. Specifically:

- *Designing out externalities*: The circular water economy optimises the amount of energy, minerals, and chemicals used in the operation of water systems in concert with other systems, optimises consumptive use of water, and uses measures or solutions which deliver the same outcome without using water.
- *Keeping resources in use*: The circular water economy aims to optimise resource yields (water use and reuse, energy, minerals, and chemicals) within water systems, optimise energy or resource extraction from the water system and maximise their reuse, and optimise value generated in the interfaces of water systems with other systems.
- *Regenerating natural capital*: The circular water economy aims to maximise environmental flows by reducing consumptive and non-consumptive uses of water, preserve and enhance natural capital (e.g. pollution prevention, quality of effluent, etc.), and ensure minimum disruption to natural water systems from human interaction and use.[11]

Water's Value in the Circular Water Economy

In addition to water being fundamental to meeting the basic needs of all living things, water in the circular water economy is valued as a service, source of energy, and a carrier of resources (Table 4.1).[12]

Table 4.1   Water's value in the circular water economy

Value	Description
Service	Water provides services including sanitation, cooling and heating, and as part of production processes
Source of energy	Water can be harnessed for hydroelectric energy, its thermal properties of absorbing thermal energy from the environment or human activity can be extracted, and bio-thermal energy can be developed from, for example, anaerobic digestion from municipal sewage
Carrier	Extracting nitrogen and phosphorous from wastewater improves the quality of outflows and reduces the cost of treatment for downstream users, reduces environmental impacts and enhances natural capital, and provides an opportunity for use as fertiliser

The Circular Water Economy Mitigating Climate Change

In the linear economy, the provision of water and the treatment of wastewater requires significant amounts of energy, the production of which is responsible for high amounts of greenhouse gas emissions. In the circular water economy, the process of water extraction, distribution, and treatment offers various options to reduce carbon emissions including increasing energy efficiency and producing renewable energy from a variety of sources including biogas from anaerobic wastewater treatment plants, heat recovery from wastewater, and use of hydropower.[13] There are many co-benefits of implementing mitigation strategies including:

- For many utilities, investments in energy-efficient technologies will have payback periods often below two years.
- Through reduced energy costs in water supply, customers can benefit from lower water tariffs.
- Lowering the energy demand of water utilities makes more energy available for other sectors.
- Reducing leakage leads to increased water availability, lower operational costs for utilities, and the protection of water resources.[14]

The Circular Water Economy Enhancing Resilience to Climate Change

The contribution of the circular water economy to climate change policy goes beyond just helping reduce greenhouse gas emissions. The concept can contribute towards making society more climate resilient, which in turn ensures development is sustainable environmentally, economically, and socially. For communities to be resilient to climate change, where resilience has been defined by the Water Services Regulation Authority (Ofwat) as *"the ability to cope with, and recover from, disruption, and anticipate trends and variability in order to maintain services for people and protect the natural environment, now and in the future"*, the water system, which comprises the physical and technological infrastructure and users, survives shocks and stresses, the people and organisations can accommodate these stresses in their day-to-day decisions, and institutional structures continue to support the capacity of people and organisations to fulfil their aims.[15,16,17,18]

Elements of Resilience

From the Ofwat description of resilience, it can be said that there are three generalisable elements of resilience in a society: systems, agents, and institutions.

Systems

Populations require high levels of infrastructure to deliver essential services, for example, water supplies. They are also linked at multiple scales to other systems, such as populations relying on ecosystem services beyond their region for flood protection. When systems fail, they jeopardise human well-being in all affected areas and hamper economic activity until their function is restored. These systems include water supplies and the ecosystems that support these. Resilient systems differ from an engineering approach to robust systems, which rely on hard protective structures, for example, large-scale water diversion schemes. Resilient systems, in contrast, ensure their functionality is retained and can be rapidly reinstated through system linkages despite some failures or operational disruptions. Factors that contribute to the resilience of systems are summarised in Table 4.2.

Agents

Agents, or actors, including individuals, households, and private and public-sector organisations have differentiated interests and can change their behaviour based on strategy, experience, and learning. Many agents are dependent on systems but are not proactively involved in their creation, management or operation of those systems while other agents are directly concerned with the management of these systems. Resilience is not evenly spread across individuals and households with poverty, gender, ethnicity, and age all contributing to differing levels of vulnerability of social groups to climate hazards through quality of housing and location and access to services and social networks. For individuals and households, their capacity to be resilient to climate hazards is determined by access to financial assets (wealth or access to credit), physical assets (house, possessions), natural assets (land), social assets (family), and human assets (healthy and skills). With climate change-related hazards typically eroding multiple types of assets, furthering impoverishing vulnerable groups, the role of local government and community organisations is to organise, plan, and coordinate disaster preparedness and

Table 4.2 Elements and factors that determine resilience

Element	Factors	Description
Systems	Flexibility and diversity	The ability to perform essential tasks under a wide range of conditions, and to convert assets or modify structures to introduce new ways of doing so. A resilient system has key assets and functions physically distributed so that one event does not affect them all at any one time (spatial diversity) and has multiple ways of meeting a given need (functional diversity)
	Redundancy, modularity	Spare capacity for contingency situations, to accommodate increasing or extreme surge pressures or demand; multiple pathways and a variety of options for service delivery; or inter-acting components composed of similar parts that can replace each other if one, or many, fail. Redundancy is also supported by the presence of buffer stocks within the system that can compensate if flows are disrupted (e.g. local water supplies to buffer imports)
	Safe failure	Ability to absorb sudden shocks (including those that exceed design thresholds) or the cumulative effects of slow-onset stress in ways that avoid catastrophic failure. Safe failure also refers to interdependence of various systems, which support each other, meaning that failure in one structure or linkage will unlikely result in cascading impacts across other systems
Agents	Responsiveness	The capacity to organise and reorganise in a timely, beneficial fashion; ability to identify problems, anticipate, plan, and prepare for a disruptive event or organisational failure; and to respond quickly
	Resourcefulness	The capacity to organise various assets and resources to act. It includes the ability to access financial and other assets, including those of other agents and systems through collaboration
	Capacity to learn	The ability to learn from past experiences, avoid repeated failures, and innovate to improve performance, as well as learn new skills

(continued)

Table 4.2 (continued)

Element	Factors	Description
Institutions	Rights and entitlements linked to system access	Rights and entitlements to use resources or access systems should be clear. Institutions that constrain the rights and entitlements can limit access to systems and services and therefore reduce resilience of vulnerable groups
	Decision-making processes	Decision-making processes should follow widely accepted principles of good governance that include transparency, accountability, and responsiveness. This includes recognition of groups most affected and ensuring they have legitimate inputs to decision-making
	Information flows	Households, businesses, community organisations, and other decision-making agents should have access to credible and meaningful information to enable judgements to be made about risk and vulnerability
	Application of new knowledge	Institutions that facilitate the generation, exchange, and application of new knowledge enhance resilience

emergency responses. These high-capacity agents have the ability to access the resources of supporting systems, including the ability to access resources provided by other agents. In addition, the capacity of individuals and organisations to learn is a critical aspect of resilience, where learning not only includes the creating and sharing of knowledge but also includes basic literacy and access to education. Factors that determine the resilience of agents are summarised in Table 4.2.

Institutions

Institutions are social rules or conventions that structure human behaviour including social and economic interactions. They can be formal or informal, overt or implicit, and created to reduce uncertainty, maintain continuity of social patterns and social order, and to stabilise human interaction in a more predictable manner. Institutions condition the ways that agents and systems interact and respond to climatic hazards. Institutions can enable or constrain individuals to organise or engage in decision-making and determine the standards to which systems are designed and managed. Institutions can enable and support, or constrain, vulnerable sectors of society. Factors that determine the resilience of institutions are summarised in Table 4.2.[19,20]

Resilience Planning

Resilience planning, according to the Asian Development Bank, is the *"process of bringing together technical, scientific, and local knowledge into decision-making processes"*. The aim of resilience planning is to build iterative, inclusive, and integrated processes to reduce the uncertainty and complexity of climate change. As part of the process, engaging multiple stakeholders needs to be more than a one-off event. Instead, engagement should be part of a cycle of action and reflection that progressively builds up capacity and understanding over time. This iterative process can increase the capacity of decision-makers including businesses, communities, households, and governments to incorporate new information and uncertainty into future plans and actions.[21]

The 3Rs in the Circular Water Economy

The development of the circular water economy, which mitigates emissions and enhances resilience to climate change, can be guided by a modified 3R approach that maximises water's value as a provider of services, source of energy, and carrier of resources that can be recovered. Specifically:

- *Reduce*: Driving continuous improvements through water conservation and water use efficiency and management.
- *Reuse and recycle*: Reuse of water within a single process or the use of harvested water for another purpose without treatment. Recycle is defined as the use of harvested water for another purpose, after treatment.
- *Recover*: There are a variety of opportunities to recover resources, including energy and nutrients, when providing water and wastewater services, all of which can provide additional revenue streams for utilities and mitigation of emissions.[22,23,24,25]

Opportunities to Implement the 3Rs

There are multiple opportunities to implement the 3Rs in the transition towards the circular water economy, including in:

- *Communities*: Water-wise communities include informed citizens who realise the role they have in conserving water and associated resources.
- *Industry*: As large water users, industry can bring circular economy solutions to scale. Increasing awareness of environmental risk means industry are increasingly seeking ways to reduce their water footprint and minimise environmental degradation.
- *Wastewater treatment plants*: In the linear economy, wastewater treatment plants remove pollutants, however, in the circular water economy, they become resource factories, energy generators, and water refineries that reuse and recover resources.
- *Drinking water treatment plants*: Drinking water treatment plants should be designed to process the same water quality time and time again with greater efficiency.
- *Agriculture*: Agriculture is the largest water user and contributes significantly to water quality, therefore, there are many opportunities for efficiencies, improvements, and value-added, competitive products and services.
- *Natural environment*: The role of the natural environment in providing services is understood but undervalued. Nature-based solutions in providing treatment, storage, buffer, and recreational solutions need to be utilised.
- *Energy generation*: Developing the circular water economy requires establishing energy independence, using less carbon-based energy, and contributing renewable energy to the grid.[26]

Technology Policies to Encourage the 3Rs

Water utilities can implement a variety of technology policies to encourage the development of the 3Rs in the circular water economy that not only mitigates greenhouse gas emissions but also enhances resilience to climate change. The term technology can be classified as hardware (physical infrastructure and technical equipment on the ground), software ('soft technologies' including approaches, processes, and methodologies including planning and decision support systems, models, knowledge transfer, and building capacity), and 'orgware' (organisation technologies for instance, organisation, ownership, and institutional arrangements).[27]

NOTES

1. WRAP, "Wrap and the Circular Economy," http://www.wrap.org.uk/about-us/about/wrap-and-circular-economy.
2. EEA, "Resource Efficient Green Economy and Eu Policies," (2014), http://www.eea.europa.eu/publications/resourceefficient-green-economy-and-eu.
3. R. C. Brears, *Natural Resource Management and the Circular Economy* (Cham, Switzerland: Springer, 2018).
4. OECD, "Material Resources, Productivity and the Environment," *OECD Green Growth Studies* (2015), http://www.oecd.org/env/waste/material-resources-productivity-and-the-environment-9789264190504-en.htm.
5. Edward B. Barbier, "The Role of Natural Resources in Economic Development," *Australian Economic Papers* 42, no. 2 (2003).
6. Goodwin Neva, Jonathan Harris, Julie A. Nelson, Brian Roach, and Mariano Torras, *Microeconomics in Context* (Armonk, NY: M.E. Sharpe, 2013).
7. World Economic Forum, "Towards the Circular Economy: Accelerating the Scale-Up Across Global Supply Chains," (2014), http://www3.weforum.org/docs/WEF_ENV_TowardsCircularEconomy_Report_2014.pdf.
8. OECD, "Material Resources, Productivity and the Environment".
9. UNEP, "Resource Efficiency: Potential and Economic Implications," (2016), http://www.resourcepanel.org/sites/default/files/documents/document/media/resource_efficiency_report_march_2017_web_res.pdf.
10. OECD, "Material Resources, Productivity and the Environment".
11. Ellen Macarthur Foundation, "Water and Circular Economy," (2018), https://www.ellenmacarthurfoundation.org/assets/downloads/ce100/Water-and-Circular-Economy-White-paper-WIP-2018-04-13.pdf.
12. Ibid.
13. GIZ, "Climate Change Mitigation in the Water Sector," (2012), https://wocatpedia.net/images/9/9e/00_GIZ_Climate_Change_Mitigation_in_the_Water_Sector.pdf.
14. Ibid.
15. Ofwat, "Towards Resilience: How We Will Embed Resilience in Our Work," (2015), https://064f1d25f5a6fb0868ac-0df48efcb31bcf2ed0366d316cab9ab8.ssl.cf3.rackcdn.com/wp-content/uploads/2015/07/pap_pos20151210towardsresiliencerev.pdf.

16. R. C. Brears, *Urban Water Security* (Chichester, UK; Hoboken, NJ: Wiley, 2016).

17. Ofwat, "Towards Water 2020—Meeting the Challenges for Water and Wastewater Services in England and Wales".

18. Brears, *Urban Water Security.*

19. Stephen Tyler and Marcus Moench, "A Framework for Urban Climate Resilience," *Climate and Development* 4, no. 4 (2012).

20. R. C. Brears, *Climate Resilient Water Resources Management* (Cham, Switzerland: Palgrave Macmillan, 2018).

21. ADB, "Urban Climate Change Resilience: A Synopsis," (2014), https://www.adb.org/publications/urban-climate-change-resilience-synopsis.

22. R. C. Brears, January 9, 2018, http://www.greengrowthknowledge.org/blog/creating-circular-water-economy.

23. Nikolaos Voulvoulis, "Water Reuse from a Circular Economy Perspective and Potential Risks from an Unregulated Approach," *Current Opinion in Environmental Science & Health* 2 (2018).

24. Ellen Macarthur Foundation, "Water and Circular Economy".

25. Cesar Casiano Flores et al., "Towards Circular Economy—A Wastewater Treatment Perspective, the Presa Guadalupe Case," *Management Research Review* 41, no. 5 (2018).

26. IWA, "Water Utility Pathways in a Circular Economy: Charting a Course for Sustainability," (2016), http://www.iwa-network.org/water-utility-pathways-circular-economy-charting-course-sustainability/.

27. UNEP-DTU UNEP-DHI Partnership, CTCN, "Climate Change Adaptation Technologies for Water: A Practitioner's Guide to Adaptation Technologies for Increased Water Sector Resilience," (2017), https://www.ctc-n.org/resources/climate-change-adaptation-technologies-water-practitioner-s-guide-adaptation-technologies.

REFERENCES

ADB. "Urban Climate Change Resilience: A Synopsis". (2014). https://www.adb.org/publications/urban-climate-change-resilience-synopsis.

Barbier, Edward B. "The Role of Natural Resources in Economic Development." *Australian Economic Papers* 42, no. 2 (2003): 253–72.

Brears, R. C. *Climate Resilient Water Resources Management.* Cham, Switzerland: Palgrave Macmillan, 2018.

———. "Creating the Circular Water Economy." Green Growth Knowledge Platform Insights Blog, 2018.

———. *Natural Resource Management and the Circular Economy.* Cham, Switzerland: Springer, 2018.

———. *Urban Water Security.* Chichester, UK; Hoboken, NJ: Wiley, 2016.

Casiano Flores, Cesar, Hans Bressers, Carina Gutierrez, and Cheryl de Boer. "Towards Circular Economy—A Wastewater Treatment Perspective, the Presa Guadalupe Case." *Management Research Review* 41, no. 5 (2018, May 21): 554–71.

EEA. "Resource-Efficient Green Economy and Eu Policies." (2014). http://www.eea.europa.eu/publications/resourceefficient-green-economy-and-eu.

Ellen Macarthur Foundation. "Water and Circular Economy."(2018). https://www.ellenmacarthurfoundation.org/assets/downloads/ce100/Water-and-Circular-Economy-White-paper-WIP-2018-04-13.pdf.

GIZ. "Climate Change Mitigation in the Water Sector." (2012). https://wocat-pedia.net/images/9/9e/00_GIZ_Climate_Change_Mitigation_in_the_Water_Sector.pdf.

Goodwin, Neva, Jonathan Harris, Julie A. Nelson, Brian Roach, and Mariano Torras. *Microeconomics in Context*. Armonk, NY: M.E. Sharpe, 2013.

IWA. "Water Utility Pathways in a Circular Economy: Charting a Course for Sustainability." (2016). http://www.iwa-network.org/water-utility-pathways-circular-economy-charting-course-sustainability/.

OECD. "Material Resources, Productivity and the Environment." *OECD Green Growth Studies* (2015). http://www.oecd.org/env/waste/material-resources-productivity-and-the-environment-9789264190504-en.htm.

Ofwat. "Towards Resilience: How We Will Embed Resilience in Our Work." (2015). https://064f1d25f5a6fb0868ac-0df48efcb31bcf2ed0366d316cab9ab8.ssl.cf3.rackcdn.com/wp-content/uploads/2015/07/pap_pos20151210 towardsresiliencerev.pdf.

———. "Towards Water 2020—Meeting the Challenges for Water and Wastewater Services in England and Wales." (2015). https://www.ofwat.gov.uk/wp-content/uploads/2015/10/pap_tec201507challenges.pdf.

Tyler, Stephen, and Marcus Moench. "A Framework for Urban Climate Resilience." *Climate and Development* 4, no. 4 (2012, October 1): 311–26.

UNEP. "Resource Efficiency: Potential and Economic Implications." (2016). http://www.resourcepanel.org/sites/default/files/documents/document/media/resource_efficiency_report_march_2017_web_res.pdf.

UNEP-DHI Partnership, UNEP-DTU, CTCN. "Climate Change Adaptation Technologies for Water: A Practitioner's Guide to Adaptation Technologies for Increased Water Sector Resilience." (2017). https://www.ctc-n.org/resources/climate-change-adaptation-technologies-water-practitioner-s-guide-adaptation-technologies.

Voulvoulis, Nikolaos. "Water Reuse from a Circular Economy Perspective and Potential Risks from an Unregulated Approach." *Current Opinion in Environmental Science & Health* 2 (2018, April 1): 32–45.

World Economic Forum. "Towards the Circular Economy: Accelerating the Scale-Up Across Global Supply Chains." (2014). http://www3.weforum.org/docs/WEF_ENV_TowardsCircularEconomy_Report_2014.pdf.

WRAP. "WRAP and the Circular Economy." http://www.wrap.org.uk/about-us/about/wrap-and-circular-economy.

CHAPTER 5

Developing the Circular Water Economy: Reduce

Abstract In the circular water economy, the concept of reduce is achieved through water conservation and water efficiency measures, in particular using less water and, where we do need to use water, making sure water is used efficiently. At the same time, reducing water usage can restore the natural environment.

Keywords Water conservation · Water efficiency · Demand management

INTRODUCTION

In the circular water economy, the concept of reduce is achieved through water conservation and water efficiency measures: using less and, where we do need to use water, making sure water is used efficiently. Water conservation is essentially doing fewer things that use water and usually involves people changing their behaviour. Water conservation is important when water supplies become unusually low, such as during severely dry summers or after natural events that disrupt water supplies. Water efficiency is when new hardware or management techniques are used to get the same level of benefits from using water, with water users needing less water in order to receive those benefits. This chapter will first discuss the implementation of water conservation and water efficiency across a

R. C. Brears, *Developing the Circular Water Economy*,
Palgrave Studies in Climate Resilient Societies,
https://doi.org/10.1007/978-3-030-32575-6_5

variety of sectors before discussing a range of technology policies that encourage reducing pressure on water supplies in the development of the circular water economy.

WATER CONSERVATION AND WATER EFFICIENCY

Water conservation and water efficiency are implemented across a variety of sectors in the development of the circular water economy. The benefits of reducing water consumption and enhancing water efficiency in the circular water economy include lower water and energy bills, lower operational and maintenance costs from water utilities having to treat less water, less energy required for pumping water, reduced carbon emissions from lower energy usage, increased water flows for natural habitats, less polluted runoff into rivers, streams, and groundwater supplies, and enhanced resilience to drought.[1,2,3,4]

Urban Areas

Reducing urban water consumption and improving water efficiency is a key solution to short-term and long-term water challenges including drought, unsustainable groundwater use, and competition over limited supplies. For instance, in drought-stricken California, residents can improve their home water efficiency by 40–60% by repairing leaks, installing the most efficient appliances and fixtures, and replacing lawns and other water-intensive landscaping with plants requiring less water. Around the world, lowering water consumption and improving water efficiency in the home can decrease water usage and lower energy bills as toilets account for almost a third of an average home's indoor water consumption, the typical family of four can use around 30% of its water on maintaining the outdoor landscape, and water heating can account for nearly a quarter of the energy consumed in a home.[5,6,7]

Case: City of Santa Cruz's Plumbing Fixture Retrofit Regulations

In the City of Santa Cruz, all residential, as well as commercial and industrial, buildings receiving water from the city are required to be retrofitted with high-efficiency plumbing fixtures at the time of sale of the property. Under City and County regulations, the seller must:

- Replace any toilets, urinals, and showerheads that do not meet the following low consumption standards:
  - Toilets: 1.28 gallons per flush (existing 1.6-gallon toilets on the property do not need to be retrofitted)
  - Urinals: 0.5 gallons per flush
  - Showerheads: 2 gallons per minute
- Arrange for inspection by either a city Water Conservation Representative or a licenced plumber or general building contractor, and
- Obtain a Water Conservation Certificate from the Water Conservation Office that certifies these changes have been made.[8]

Industry

In industrial facilities, water is used in a wide range of activities including incorporation in the final product, washing or rinsing of raw materials, intermediates, or final products, preparation of solvents or slurries, cleaning equipment and space, removing or providing heat, meeting hygienic and domestic needs, and irrigation of landscape space. Around the world, industrial facilities that have systematically adopted water efficiency measures have reduced water consumption by 20–50%, and up to 90% if advanced measures are implemented. Table 5.1 lists a range of industrial efficiency measures and their associated water-saving potential.[9]

There are a number of different options for lowering water consumption and improving water efficiency in industry in order of increased complexity in process changes, including:

Table 5.1 Water-saving potential in industry

Efficiency measures	Potential savings (%)
Closed loop reuse	~90
Closed loop recycling with treatment	~60
Automatic shut-off valves	~15
Counter-current rinsing	~40
High-pressure, low-volume upgrades	~20
Reuse of wash water	~50

- *Improved production planning and sequencing*: Re-adjusting the production plans with a focus on minimising water consumption.
- *Good housekeeping*: Introducing more sensible and more resource-conscious routines in operations.
- *Process/equipment modifications*: Making modifications in processes or equipment, with relevant retrofits, if necessary.
- *Product/material changes*: Changing feedstocks used in production or designing completely new products that lead to reduced water demand and/or less effluent generation.
- *Replacing equipment/technology*: Substituting existing technologies with more effective and efficient ones.[10]

Case: Welsh Water Creating Water-Efficient Businesses

To help business customers reduce water wastage and lower overheads by lowering water and energy bills, Welsh Water offers business customers two types of water efficiency audits:

- *Standard water efficiency audit*: This is an assessment of 'domestic' water use on-site (toilets, taps, kitchens, and cleaning facilities). The aim of this audit is to identify opportunities for saving water and to quantify this in volumetric terms. A utility representative will review the customer's current water consumption and help identify areas where efficiency gains can be made. Specifically, the audit entails:
  - Making a general inspection of domestic water systems
  - Analysing meter readings to determine current consumption patterns
  - Identifying site sub-metering opportunities
  - Assessing the extent of leakage on-site
  - Assessing urinal cistern control mechanisms
  - Commenting on water management options
  - Summarising total water saving potential.
- *Process audit*: If a business uses water in its processes, a process audit can help identify efficiencies. The process audit is carried out by a Welsh Water industrial process engineer who will look at water consumption within the production and operational processes found on-site. The audit can be scaled to reflect the needs of each customer, but the key parts of the audit service include:

- *Site walkover*: An experienced engineer will carry out a detailed observation of on-site processes, identifying areas of high consumption or wastage, plus cost-saving opportunities.
- *Data analysis*: A detailed site data analysis will be taken including efficiency correlation analysis, analysis of production, water, and effluent volumes, and, where appropriate, concentrations.
- *Mass balance*: A site-wide balance will be created that shows unidentified losses and the costs of those losses.
- *Cost-benefit analysis*: This will identify opportunities for cost-savings, showing pay-back periods and priorities by profitability.[11]

Buildings

Residential as well as institutional and commercial buildings, such as schools, hospitals, shopping centres, offices, hotels, and restaurants, are places where significant amounts of water are consumed and where considerable savings of water can be captured. Water use in buildings is projected to increase with urbanisation, rising living standards, and the growing service sector in national economies. In the United States, it is estimated that buildings use more than 400 billion gallons of water per day and that in the federal sector alone expenditures for water and sewer services reach up to $1 billion annually and that moderate gains in water efficiency could save as much as $240 million per year: in fact, if water savings of 40% were achieved it would be enough water to supply a population of around 1.8 million.[12,13] There are a number of strategies that can be employed to reduce the amount of water consumed in a building including system optimisation (i.e. efficient water system design, leak detection, and repair), water conservation measures, and water reuse/recycling systems.

Case: New York City Water Challenge to Restaurants

In 2015, New York City's Department of Environmental Protection announced the New York City Water Conservation Challenge to Restaurants. Thirty restaurants joined the challenge with each restaurant working to reduce their annual water consumption by 5%

and thereby aim to save a total of around three million gallons of water each year. The challenge provided structured resources for participants through which they could:

- Track and benchmark their water use
- Attend four workshops featuring green hospitality experts
- Develop a Water Conservation Plan, which contains the restaurant's water-saving strategies for the year of the challenge
- Actively implement a Water Conservation Plan over the year of the challenge and beyond.[14]

Agriculture

There are a variety of options available for reducing water consumption and improving water efficiency in agriculture including efficient irrigation technologies, improved irrigation scheduling, regulated deficit irrigation, and practices to enhance soil moisture. Weather-based irrigation scheduling uses data about local weather conditions to determine how much water a crop needs, regulated deficit irrigation imposes water stress on certain crops that have drought-tolerant life stages, while sprinkler and drip irrigation systems usually have higher distribution uniformities and water-use efficiencies than traditional flood or gravity irrigation systems, with drip irrigation systems slowly releasing low-pressure water from plastic tubing placed near the plant's root zone, allowing for the precise application of water and fertiliser to meet crop needs.[15] Other practices to lower water consumption and improve water efficiency are listed in Table 5.2.[16]

Table 5.2 Practices to lower water consumption and improve water efficiency in agriculture

Practice	Description
Dry farming	Farming techniques that use mulches, residue management in crop fields, etc. help retain soil moisture for crop production
Rotational grazing	Animals are shifted between fields to promote pasture regrowth. Good grazing management increases the field's water absorption and decreases runoff, making pastures more drought resistant. Increased soil organic matter and better forage cover are also water-saving benefits of rotational grazing

(continued)

Table 5.2   (continued)

Practice	Description
Compost and mulch	Compost, or decomposed organic matter used as fertiliser, enhances water-holding capacity and improves soil structure. Mulch is a material spread on top of the soil to retain soil moisture and can be made from organic materials that can break down into compost
Cover crops	These protect soil that would otherwise go bare. They reduce weeds and increase soil fertility and organic matter. Cover crops allow water to penetrate the soil more easily which improves its water-holding capacity
Conservation tillage	This involves specialised ploughs or other equipment to partially till the soil but leaves at least 30% of vegetative crop residue on the surface, helping increase water absorption and reduce evaporation
Going organic	Organic methods help retain soil moisture and healthy soil is rich in organic matter and microbial life which serves as a sponge that absorbs and retains moisture for plants. In addition, it reduces toxic pesticides from entering waterways

> **Case: New York State New Farmers Grant Fund Program**
>
> New York State has allocated $1 million in the 2018–2019 state budget to support beginning farmers who have chosen farming as a career and who materially and substantially participate in the production of an agricultural product on their farm. The grant fund will help a farmer improve farm profitability including advancing innovative techniques that increase sustainable agricultural production practices such as organic farming and reduction of farm water use etc. Eligible farmers may receive a minimum of $15,000 and a maximum of $50,000 for up to 50% of total project costs. The remaining 50% must be matched by the recipient.[17]

REDUCING WATER WASTAGE AT THE POINT OF CONSUMPTION

Demand management promotes water conservation and water efficiency during both normal and abnormal conditions, through changes in practices, culture, and people's attitudes towards water resources. Demand management seeks to reduce the loss and misuse of water, optimise the use of water, and facilitate major financial and infrastructural savings by minimising the need to meet increasing demand with new water supplies. The benefits of demand management include reduced water and electricity bills, reduced carbon emissions from pumping and heating water, reduced leakage, and more water for a healthier environment.[18] There are a variety of demand management tools available to water utilities to promote water conservation and water efficiency including the following:

Pricing of Water

The main pricing structures used by water utilities to promote the conservation of water resources are volumetric rates, increasing block tariffs, and two-part tariff systems. A volumetric rate is a charge based on the volume used at a constant rate. Therefore, the amount users pay for water is strictly based on the amount of water consumed. An increasing block tariff contains different prices for two or more pre-specified quantities (blocks) of water, with the price increasing with each successive block. The pricing of water can also be done using a two-part tariff system: a fixed and a variable component. In the fixed component, water users pay one amount independently of consumption and this covers infrastructural and administrative costs of supplying water. Meanwhile, the variable amount is based on the quantity of water consumed and covers the costs of providing water as well as encouraging conservation.[19]

Case: Queensland Urban Utilities' Prices and Charges

Queensland Urban Water Utilities' water charges incorporate a variety of fixed and variable costs (listed in Table 5.3):

- *State bulk water price*: Covers the cost of the treated water bought from the State Government.
- *Queensland Urban Utilities* distributor-*retailer price*: The charge to maintain water quality and to deliver water to properties. Water usage is charged per kilolitre (1 kL = 1000 litres) and is based on water meter reads.
- *Water services*: This helps pay for water services, including the installation, maintenance, and upgrade of infrastructure used to treat and transport clean drinking water to properties.[20]

Table 5.3 Queensland Urban Utilities' prices and charges for customers

Water usage		
	Residential (per kilolitre)	Business (per kilolitre)
State bulk water price	$3.017	$3.017
Queensland Urban Utilities distributor-retailer price		
Tier 1 (up to 74 kL per quarter)	$0.793	$0.793
Tier 2 (over 74 kL per quarter)	$1.569	$1.659
Water services		
	Residential (per quarter)	Business (per quarter)
Queensland Urban Utilities water service charge	$56.13	$58.77

Water Meters

Before water users can be charged for the amount of water consumed, the dwelling or building must have water meters to measure the volume of water consumed. Automated meter readers (AMRs) are 'one-way' readers that send usage data back to the utility. In comparison, advanced metering infrastructure (AMI) is a 'two-way' solution that creates a network between the meters and the utility's information system. Data flows both ways facilitating not only remote meter reading but also the use of variable pricing.[21]

Case: City of Guelph's Residential Sub-water Meter Program

The City of Guelph is providing rebates towards the purchase and installation of sub-meters for residential buildings including rental apartment buildings, condominiums, and single-family homes. Using sub-meters to monitor water use helps notify property owners and managers if individual units have high water demand, which could indicate leaks, inefficient plumbing products, wasteful water use habits, etc. The City of Guelph will rebate $125 for each permanently installed sub-water meter. The city will also provide a rebate of $100 for each add-on sub-water meter with smart technology.[22]

Incentive Schemes

Economic instruments such as subsidies or rebates are used to modify water users' behaviour in a predictable, cost-effective way, that is, reduce wastage and lower water consumption. Subsidies (incentives) are commonly used to encourage the uptake of water-saving devices or water-efficient appliances as positive incentives are found to be more effective than disincentives in promoting water conservation. In addition, incentives have been found to reduce the gap between the time the incentive is presented and behavioural change as compared to disincentives. To accelerate the replacement of old water-using fixtures, water utilities commonly offer rebates to customers who purchase water-efficient toilets, taps, and showerheads.[23]

Case: City of Calgary's Multi-unit Toilet Replacement Programme

For owners or managers of buildings with three or more suites, the City of Calgary is offering a $50 Toilet Replacement Rebate to replace old toilets with new high-efficiency WaterSense[TM] labelled toilets that can expect to use 30% less water. Over two years, the total savings on the water bill will have covered the cost of the new high-efficiency toilets. To receive the rebate, owners or managers of multi-unit buildings must first have removed the old 13 litres or more toilet(s) and have purchased and installed high-efficiency, WaterSense[TM] labelled toilet(s).[24]

Regulations

Water management generally come in the form of temporary and permanent regulations:

- *Temporary regulations*: Restrict certain types of water use during specified times and/or restrict the level of water use to a specified amount. These programmes are usually enacted during times of severe water shortages and cease once the shortage has passed. Restrictions are placed on non-essential water uses.
- *Permanent regulations*: Amendments to building codes or ordinances requiring the installation of water-saving devices and maximum water use standards for plumbing fixtures.[25]

Case: City of Vancouver's Water Restrictions

The City of Vancouver applies water restrictions on the use of treated drinking water over the period of May 1 to October 15. If water is used outside of this time, a fine of $250 could be imposed. Year-round, automatic shut-off devices are required on hoses for all watering and washing activities regardless of the current level of water restrictions. Some of the main restrictions are listed in Table 5.4.[26]

Table 5.4   City of Vancouver water restriction examples

Activity	Stage 1	Stage 2	Stage 3	Stage 4
Residential lawn watering	Restricted **Even-numbered addresses:** Wednesday and Saturday 4 a.m.–9 a.m. ONLY **Odd-numbered addresses:** Thursday and Sunday 4 a.m.–9 a.m. ONLY	Restricted **Even-numbered addresses:** Wednesday 4 a.m.–9 a.m. ONLY **Odd-numbered addresses:** Thursday 4 a.m.–9 a.m. ONLY	Prohibited No lawn watering allowed	Prohibited No lawn watering allowed
Non-residential lawn watering	Restricted **Even-numbered addresses:** Monday 1 a.m.–6 a.m. Friday 4 a.m.–9 a.m. **Odd-numbered addresses:** Tuesday 1 a.m.–6 a.m. Friday 4 a.m.–9 a.m.	Restricted **Even-numbered addresses:** Monday 1 a.m.–6 a.m. ONLY **Odd-numbered addresses:** Tuesday 1 a.m.–6 a.m. ONLY	Prohibited No lawn watering allowed	Prohibited No lawn watering allowed
Surface and power washing	Allowed	Restricted For safety or painting prep ONLY Commercial companies exempt	Restricted For safety or painting prep ONLY Commercial companies exempt	Prohibited Not allowed unless ordered by a regulatory authority

Water Efficiency Labelling

The labelling of household appliances according to water efficiency is important in reducing household water consumption by eliminating unsustainable products from the market; however, this is provided the labelling scheme is clear and comprehensible and identifies both private and public benefits of conserving water. Nonetheless, people are more likely to respond to eco-labels if the environmental benefits match closely personal benefits such as reduced water bills.[27]

Case: Hong Kong's Water Efficiency Labelling Scheme

Since 2009, Hong Kong has been running its Voluntary Water Efficiency Labelling Scheme (WELS) to cover the common types of plumbing fixtures and water-consuming devices. Products participating in WELS incorporate a water efficiency label that informs consumers the level of water consumption and water efficiency (WELS Labels display the grade of water efficiency with Grade 1 the most water efficient and Grade 4 the least), helping consumers choose water efficient products for water conservation. On February 1st, 2017, the government launched the mandatory use of devices registered under WELS to further enhance water efficiency. Under the mandatory requirements, for plumbing works for kitchens of domestic premises as well as for bathrooms and toilets in all premises that involve the installation of showers for bathing, water taps, urinal flushing valves, etc., models of devices registered under WELS are to be used.[28,29]

Consumer Education and Awareness

Education of the public is crucial in generating an understanding of water scarcity and creating acceptance of the need to implement water conservation programmes. Water utilities can promote water conservation in schools to increase young people's knowledge of the water cycle and encourage the sustainable use of scarce water resources. Meanwhile, water utilities can raise public awareness of the need to conserve water resources. This can be done through many formats including:

- *Public information*: Printed literature distributed or available for the general public, public service announcements and advertisements on billboards and public transportation, television commercials, newspaper articles and advertisements, Internet and social media campaigns.
- *Public events*: Customers can receive information on water conservation tips and receive water-saving devices at conservation workshops, expos, fairs, etc. as people frequently make poor choices with regard to environmentally friendly products or services due to misinformation or lack of information.
- *Information in water bills*: Water bills should be understandable enabling customers to easily identify volume of usage, rates, and charges. Water bills should be informative enabling customers to compare their current bill with previous bills. Finally, water bills should contain water conservation tips to help customers make informed decisions on future water use.[30]

Case: South East Water's Online Games

South East Water in Melbourne, Australia, provides several online games for students to learn about the natural and urban water cycles. These are summarised in Table 5.5.[31]

Table 5.5 South East Water's online games

Game	Description	Activity
Melbourne water cycle game	A lot of work goes into Melbourne's water cycle: catching clean water to have on tap, getting it to homes, and then taking away the sewage and cleaning it	Try your hand at running Melbourne's water cycle and see how long it takes you to get the water from the reservoir to the ocean
Natural water cycle game	All ages will enjoy learning about or testing their knowledge of the natural water cycle in this engaging game	Help Whoosh bounce warmth from the sun into the ocean to start the water cycle. Move to the next level to melt the snow and return the water to the ocean. How long will it take you to make evaporation, condensation, and precipitation?
Water sources game	Take on the challenge to learn where different types of water come from including rain, grey, recycled, and desalination	Help Whoosh connect the pipes and join the sources. Look out for all the obstacles along the way!

WATER UTILITY-LED WATER CONSERVATION AND WATER EFFICIENCY MEASURES

Water utilities can reduce leakage and use natural infrastructure to enhance water conservation and lower operating costs of treating wastewater.

Reducing Leakage

Non-revenue water (NRW) is the difference between the amount of water a water utility pumps into the distribution system and the amount of water that is billed to its consumers. The difference is commonly known as 'leakage' and in many cities it can be 25–50% of distributed water. NRW comprises:

- *Apparent losses*: Also termed 'commercial losses', this is caused by inaccurate metering, data handling errors, and illegal tapping.
- *Real losses*: Also termed 'physical losses', this comprises leakage from all parts of the system and overflows at storage tanks. Real losses are caused by poor operations and maintenance and poor quality of underground assets.
- *Unbilled authorised consumption*: This is water used for flushing, firefighting, and water provided for free to certain consumer groups.

Factors causing this gap include inaccurate billing systems, deficient customer registration, leakage caused by deteriorating infrastructure, poor water pressure management, inaccurate metering, reservoir overflow, unnecessary flushing, insufficient management, and illegal connections to the water network. A NRW programme that reduces water loss and increases revenue will also provide a variety of additional benefits for the utility and its consumers, including reduced stress on the area's water resources, reduced energy consumption for abstraction, treatment, and distribution, and improved water quality due to optimised water distribution as chlorine content in the distributed water will be better controlled and the risk of pollution related to burst and periods with low pressure or vacuum will be reduced.[32]

Case: Severn Trent Detecting Leaks from Space

Severn Trent in the United Kingdom is aiming to reduce leakage levels by 15%. In addition to using drones to spot leaks, the utility is conducting two trials to find leaks in different ways. In the first trial, Severn Trent is working with Rezatec, who analyse satellite data and imagery to monitor changes in the landscape near to a pipe. This includes changes to the vegetation, water content in soil, water accumulation, and sub-centimetre ground movement, all of which helps identify potential leakage. In the other trial, the utility is working with Utilis, using a Japanese satellite, to detect the signature of longwave radiation of chlorine in water, tracking the 'signature' of drinking water escaping into the ground. The trials have to date found points of interest, with teams then following up with onsite investigations.[33]

Implementing Natural Infrastructure

Natural infrastructure uses landscape management strategies including restoration and conservation to provide essential services such as clean and abundant water supply, aquifer recharge, and flood control. In the context of the circular water economy, water utilities can use natural infrastructure, such as riparian buffers, reservoir and catchment restoration, and constructed wetlands, to enhance water conservation, restore and enhance the natural environment, as well as reduce operating expenditure related to treatment processes.[34,35]

Case: Artificial Wetlands in Trosa, Sweden

Since 2003, wastewater from the town of Trosa, which has a population of around 4500 inhabitants, has been treated in artificial wetlands, after the basic treatment in the sewage plant. Because of the wetlands, the Trosa river and town bay have been spared from eutrophicating substances (plant nutrient) as well as contagious elements (pathogens). The wetlands have an area of around six hectares and after around one week the wastewater is of high quality with bathing water quality achieved for almost the whole year. More than 50% of the nitrogen that comes with the wastewater from households is removed when the water has been treated in the treatment plant and the wetlands. The remaining amount of phosphorous after treatment is on average <0.1 mg/litre over the year and the reduction of nitrogen in the wetlands is about one tonne per annum.[36]

NOTES

1. Pacific Institute, "Multiple Benefits of Water Conservation and Efficiency for California Agriculture," https://pacinst.org/publication/multiple-benefits-of-water-conservation-and-efficiency-for-california-agriculture/.
2. Wellington City Council, "Water Conservation and Efficiency Plan 2011," (2011), https://wellington.govt.nz/~/media/your-council/plans-policies-and-bylaws/plans-and-policies/a-to-z/waterconservation/files/waterconservation.pdf?la=en.
3. R. C. Brears, *Urban Water Security* (Chichester, UK and Hoboken, NJ: Wiley, 2016).
4. *The Green Economy and the Water-Energy-Food Nexus* (London, UK: Palgrave Macmillan, 2017).
5. Pacific Institute and NRDC, "Agricultural Water Conservation and Efficiency Potential in California," (2014), http://pacinst.org/wp-content/uploads/2014/06/ca-water-ag-efficiency.pdf.
6. U.S. Department of Energy, "Guide to Home Water Efficiency," (2010), https://www.energy.gov/sites/prod/files/guide_to_home_water_efficiency.pdf.
7. Wellington Water, "Buy Water Efficient Products (Wels)," https://www.wellingtonwater.co.nz/your-water/drinking-water/looking-after-your-water/water-conservation/water-conservation-inside/buy-water-efficient-products-wels/.
8. City of Santa Cruz, "Plumbing Fixture Retrofit Regulations," http://www.cityofsantacruz.com/government/city-departments/water/conservation/regulations/plumbing-fixture-retrofit-regulations.
9. Arab Forum for Environment and Development, "Water Efficiency Handbook: Identifying Opportunities to Increase Water Use Efficiency in Industry, Buildings, and Agriculture in the Arab Countries," (2014),

http://www.afedonline.org/WEH2014/eng/WATER-Efficiency-Handbook-ENGLISH.pdf.

10. Ibid.

11. Welsh Water, "Water Efficiency Audit," https://www.dwrcymru.com/en/Business/Our-Services/Water-Efficiency-Audit.aspx.

12. U.S. Green Building Council, "Green Building 101: How Does Water Efficiency Impact a Building?"

13. National Institute of Building Sciences, "Water Conservation," https://www.wbdg.org/resources/water-conservation.

14. NYC DEP, "New York City Water Challenge to Restaurants," http://home2.nyc.gov/html/dep/html/ways_to_save_water/nyc-water-restaurants-challenge.shtml.

15. Pacific Institute and NRDC, "Agricultural Water Conservation and Efficiency Potential in California."

16. S. Barua, R. Kumar, and P. Singh, "10 Water Saving Techniques in Agriculture," http://www.indiawaterportal.org/articles/10-water-saving-techniques-agriculture.

17. Empire State Development, "New York State New Farmers Grant Fund Program," https://esd.ny.gov/new-farmers-grant-fund-program.

18. Brears, *Climate Resilient Water Resources Management* (Cham, Switzerland: Palgrave Macmillan, 2018).

19. *Urban Water Security.*

20. Queensland Urban Utilities, "Prices and Charges 2018–19," https://www.urbanutilities.com.au/residential/accounts-and-billing/prices-and-charges-2018-19.

21. Brears, *Urban Water Security.*

22. City of Guelph, "Residential Sub-water Meter Program," http://guelph.ca/living/environment/rebates/residential-sub-water-meter-program/.

23. Brears, *Urban Water Security.*

24. City of Calgary, "Multi-unit Toilet Replacement Program," http://www.calgary.ca/UEP/Water/Pages/Water-conservation/Indoor-water-conservation/Toilet-rebates/Hotel-motel-toilet/Hotel-Motel-Toilet-Replacement-Program.aspx.

25. Brears, *Urban Water Security.*

26. City of Vancouver, "New Water Restrictions Explained," https://vancouver.ca/home-property-development/understanding-watering-restrictions.aspx.

27. Brears, *Climate Resilient Water Resources Management.*

28. Water Supply Department, "Water Efficiency Labelling Scheme," https://www.wsd.gov.hk/en/plumbing-engineering/water-efficiency-labelling-scheme/index.html.

29. Government of Hong Kong, "WSD Launches Requirements for Mandatory Use of Devices Under Water Efficiency Labelling Scheme," https://www.info.gov.hk/gia/general/201701/20/P2017011900857.htm.
30. Brears, *Urban Water Security.*
31. South East Water, "Play Games and Explore," https://www.education-southeastwater.com.au/im-a-student/play-games-and-explore.
32. State of Green, "Reducing Urban Water Loss," (2016), https://stateof-green.com/en/publications/reducing-urban-water-loss/.
33. Severn Trent, "We're Detecting Leaks from Space!" https://www.stwater.co.uk/news/news-releases/wearedetectingleaksfromspace/.
34. R. C. Brears, *Blue and Green Cities: The Role of Blue-Green Infrastructure in Managing Urban Water Resources* (London, UK: Palgrave Macmillan, 2018).
35. Robert I. McDonald et al., "Estimating Watershed Degradation over the Last Century and Its Impact on Water-Treatment Costs for the World's Large Cities," *Proceedings of the National Academy of Sciences* 113, no. 32 (2016).
36. Smart City Sweden, "Artificial Wetlands in Trosa," https://smartcitysweden.com/reference-objects/77/artificial-wetlands-in-trosa/.

REFERENCES

Arab Forum for Environment and Development. "Water Efficiency Handbook: Identifying Opportunities to Increase Water Use Efficiency in Industry, Buildings, and Agriculture in the Arab Countries." (2014). http://www.afedonline.org/WEH2014/eng/WATER-Efficiency-Handbook-ENGLISH.pdf.
Barua, S., R. Kumar, and P. Singh. "10 Water Saving Techniques in Agriculture." http://www.indiawaterportal.org/articles/10-water-saving-techniques-agriculture.
Brears, R. C. *Blue and Green Cities: The Role of Blue-Green Infrastructure in Managing Urban Water Resources.* London, UK: Palgrave Macmillan, 2018.
———. *Climate Resilient Water Resources Management.* Cham, Switzerland: Palgrave Macmillan, 2018.
———. *The Green Economy and the Water-Energy-Food Nexus.* London, UK: Palgrave Macmillan, 2017.
———. *Urban Water Security.* Chichester, UK and Hoboken, NJ: Wiley, 2016.
City of Calgary. "Multi-unit Toilet Replacement Program." http://www.calgary.ca/UEP/Water/Pages/Water-conservation/Indoor-water-conservation/Toilet-rebates/Hotel-motel-toilet/Hotel-Motel-Toilet-Replacement-Program.aspx.
City of Guelph. "Residential Sub-water Meter Program." http://guelph.ca/living/environment/rebates/residential-sub-water-meter-program/.

City of Santa Cruz. "Plumbing Fixture Retrofit Regulations." http://www. cityofsantacruz.com/government/city-departments/water/conservation/ regulations/plumbing-fixture-retrofit-regulations.

City of Vancouver. "New Water Restrictions Explained." https://vancouver.ca/ home-property-development/understanding-watering-restrictions.aspx.

Empire State Development. "New York State New Farmers Grant Fund Program." https://esd.ny.gov/new-farmers-grant-fund-program.

Government of Hong Kong. "WSD Launches Requirements for Mandatory Use of Devices Under Water Efficiency Labelling Scheme." https://www.info. gov.hk/gia/general/201701/20/P2017011900857.htm.

McDonald, Robert I., Katherine F. Weber, Julie Padowski, Tim Boucher, and Daniel Shemie. "Estimating Watershed Degradation over the Last Century and Its Impact on Water-Treatment Costs for the World's Large Cities." *Proceedings of the National Academy of Sciences* 113, no. 32 (2016): 9117–22.

National Institute of Building Sciences. "Water Conservation." https://www. wbdg.org/resources/water-conservation.

NYC DEP. "New York City Water Challenge to Restaurants." http://home2. nyc.gov/html/dep/html/ways_to_save_water/nyc-water-restaurants-challenge.shtml.

Pacific Institute. "Multiple Benefits of Water Conservation and Efficiency for California Agriculture." https://pacinst.org/publication/multiple-benefits-of-water-conservation-and-efficiency-for-california-agriculture/.

Pacific Institute and NRDC. "Agricultural Water Conservation and Efficiency Potential in California." (2014). http://pacinst.org/wp-content/uploads/ 2014/06/ca-water-ag-efficiency.pdf.

Policy Research Initiative, Government of Canada. *Economic Instruments for Water Demand Management in an Integrated Water Resources Management Framework: Synthesis Report.* Policy Research Institute, 2005.

Queensland Urban Utilities. "Prices and Charges 2018–19." https://www. urbanutilities.com.au/residential/accounts-and-billing/prices-and-charges-2018-19.

Severn Trent. "We're Detecting Leaks from Space!" https://www.stwater.co.uk/ news/news-releases/wearedetectingleaksfromspace/.

Smart City Sweden. "Artificial Wetlands in Trosa." https://smartcitysweden. com/reference-objects/77/artificial-wetlands-in-trosa/.

South East Water. "Play Games and Explore." https://www.educationsoutheast-water.com.au/im-a-student/play-games-and-explore.

State of Green. "Reducing Urban Water Loss." (2016). https://stateofgreen. com/en/publications/reducing-urban-water-loss/.

U.S. Department of Energy. "Guide to Home Water Efficiency." (2010). https:// www.energy.gov/sites/prod/files/guide_to_home_water_efficiency.pdf.

U.S. Green Building Council. "Green Building 101: How Does Water Efficiency Impact a Building?"

Water Supply Department. "Water Efficiency Labelling Scheme." https://www.wsd.gov.hk/en/plumbing-engineering/water-efficiency-labelling-scheme/index.html.

Wellington City Council. "Water Conservation and Efficiency Plan 2011." (2011). https://wellington.govt.nz/~/media/your-council/plans-policies-and-bylaws/plans-and-policies/a-to-z/waterconservation/files/waterconservation.pdf?la=en.

Wellington Water. "Buy Water Efficient Products (Wels)." https://www.wellingtonwater.co.nz/your-water/drinking-water/looking-after-your-water/water-conservation/water-conservation-inside/buy-water-efficient-products-wels/.

Welsh Water. "Water Efficiency Audit." https://www.dwrcymru.com/en/Business/Our-Services/Water-Efficiency-Audit.aspx.

CHAPTER 6

Developing the Circular Water Economy: Reuse and Recycle

Abstract Reused and recycled water can be used for a variety of non-potable uses including industrial processes and irrigation of agricultural land. Recycled water can also be blended with surface or groundwater to increase supplies. This reduces the economic and environmental costs related to establishing new water supplies in the development of the circular water economy.

Keywords Water reuse · Recycled water · Water recycling

INTRODUCTION

Water reuse involves collecting, treating, and reusing wastewater without treatment and can also involve the re-use of greywater or rainwater in houses or buildings. Meanwhile recycled water, which is reclaimed water that has been treated, can be used for non-potable uses such as industrial processes and irrigation of agriculture land. If treated appropriately recycled water can be blended with surface or groundwater to increase supplies. This reduces the economic and environmental costs related to establishing new water supplies in the development of the circular water economy. This chapter will first define the terms 'water reuse' and 'water recycling' before discussing the implementation of water reuse and water recycling systems across a variety of sectors in the development of the circular water economy.

© The Author(s) 2020   65
R. C. Brears, *Developing the Circular Water Economy,*
Palgrave Studies in Climate Resilient Societies,
https://doi.org/10.1007/978-3-030-32575-6_6

REUSE AND RECYCLING DEFINED

Regarding water conservation hierarchy terminology, reuse is defined as the reuse of water within a single process or the use of harvested water for another purpose without treatment, while recycle is defined as the use of harvested water for another purpose, after treatment. Treatment can be tailored to meet the water quality requirements of a planned use.

Table 6.1 Benefits of water reuse and recycling

Benefit	Linear economy challenge	Circular water economy solution
Decreases diversion of freshwater from sensitive ecosystems	Plants, wildlife, and fish depend on sufficient water flows for healthy habitats. Lack of adequate flow, from diversion for agricultural, urban, and industrial purposes, can cause deterioration of water quality and ecosystem health	Water reuse and recycling can meet water demand, freeing up considerable amounts of water for the environment and increase flows to vital ecosystems
Decreases discharge to sensitive water bodies	Discharge of wastewater can impact ocean, estuary or stream health	Water reuse and recycling can be used to eliminate or decrease wastewater discharge into receiving water bodies, enhancing ecosystem health
May be used to create/enhance wetlands and riparian habitats	Wetlands provide multiple benefits including wildlife habitat, water quality improvement, flood diminishment, etc. However, impaired or dried streams from water diversion damages these wetlands	Water flow can be augmented with reused and recycled water to sustain and improve aquatic and wildlife habitat
Reduces and prevents pollution	Pollutant discharges to oceans, rivers, and other water bodies impact human and natural health, such as increased nutrient loading causing harmful algal blooms	Reused and recycled water can be diverted for use in other locations, such as recycled water containing higher levels of nutrients can be used for agricultural and landscape irrigation, lessening the need to apply synthetic fertilisers
Saves energy	As demand for water increases, more water is extracted, treated, and transported, often over great distances. This increases demand for energy. If the local source is groundwater, lower groundwater levels increase the energy required to pump water	Reusing and recycling water on site or nearby reduces the energy needed to distribute water. Tailoring water quality to a specific water use also reduces the energy needed to treat water

Water recycling, in general, involves the reclamation of water from wastewater for non-potable or potable use, which can be supplied back to the water system either directly or indirectly. Water for recycling can come from centralised schemes or from small on-site systems involving, for example, treated sewage or greywater. Recycled water can be used for a variety of purposes including agricultural and landscape irrigation, industrial processes, toilet flushing, and replenishing groundwater basins. In addition to providing a dependable, locally controlled, climate resilient water supply, water reuse and recycling provide a range of benefits summarised in Table 6.1.[1,2,3,4,5,6]

INDUSTRIAL WATER REUSE AND RECYCLING

Industrial water can be reused (wastewater is reused directly) or recycled (treated before reuse) within a business itself or between several businesses.

Direct Reuse Within a Business

A business can directly reuse wastewater that is clean enough for the purpose for which it is being reused. Process water is produced by industrial processes such as cooling and heating and usually contains few contaminants after use. Cooling towers are one of the most common water technologies in use by industry and is frequently used for washing processes.[7]

Direct Reuse Between Businesses

Direct wastewater reuse can be practised between businesses, with the exchange of waste product for the mutual benefit of two or more businesses known as 'industrial symbiosis'. All members benefit from industrial symbiosis by reducing the inputs necessary in their production process, i.e. water, or by reducing the costs of wastewater treatment. Industrial symbiosis can take place through the exchange of by-products, sharing the management of utilities, and sharing of ancillary services. Some examples of direct reuse of wastewater in industrial symbiosis include the exchange of process water from one business to another and subsequent reuse, reuse of organic waste or wastewater for biogas production, reuse of wastewater for aquaculture of plants or animals, etc.[8]

Case: Smart Water (Re-)Use in Flanders

Flanders, Belgium, has a high demand for water in various industries including chemical plants, power plants, refineries, steel industries, and agriculture yet has less water availability than Spain, Portugal, and Greece. With resource recovery as its focus, Flanders' 'Center for Advanced Process Technology for Urban REsource recovery' (CAPTURE) is a research and accelerator platform that unites the public and private sector, as well as academics on developing circular economy solutions for water as well as plastics and carbon dioxide. CAPTURE aims to develop new, robust, and disruptive technologies to efficiently use every drop of water. CAPTURE aims to ensure close collaboration between all stakeholders of the water cycle to create a close symbiosis enabling smart water (re-)use and the recovery of essential resources from various water flows. Specifically, CAPTURE aims to enhance the symbiosis in seven research domains: Water fit-for-use; wastewater treatment of the future; nutrient recovery; metal recovery; energy and organics from water; identification of useful resource streams; and advanced modelling and design for efficient pilot demonstrations.[9]

Treat and Reuse (Recycling)

If wastewater is not suitable for direct reuse, decentralised wastewater treatment systems may be used to reduce the level of contaminants to a level that is safe for reuse. This can be done within a business for its own reuse or between businesses. The specific treatment options are dependent on the quality required at the end: Water used for irrigation may be treated in a constructed wetland while advanced options including membrane filtration and activated carbon may produce treated wastewater of a higher quality than freshwater.[10]

Case: Custom-Designed Process Water in the Netherlands

Evides, a water utility that provides drinking water to 2.5 million consumers and businesses in three Dutch provinces of Zeeland, the south-western part of Zuid-Holland, and the south-western part of Noord-Brabant, offers industrial water, or 'process water', that is customised for specific requirements and preferences of each customer. In 2017 alone, Evides Industriewater provided 96.9 million cubic metres of industrial water and 158.7 million cubic metres of drinking water through more than 14,000 kilometres of the pipeline network. The utility focuses on providing process water to the chemical industry, petrochemicals, and food industry with an overall emphasis on tailoring water to the needs of customers. The utility delivers different process water on the basis of a Design, Build, Finance, and Operate (DBFO) contract, which involves Evides Industriewater, on behalf of its customers, purifying water for reuse. Some of the main industrial water products Evides Industriewater offers to customers include agricultural and irrigation water, cooling water, industrial water, and demineralised water. Demand has now resulted in Evides Industriewater providing solutions to 75 installations, including 12 DBFO plants as well as mobile installations in the Benelux and Germany.[11]

Benefits of Industrial Water Reuse and Recycling

There are many benefits of industrial water reuse and recycling including reduced freshwater costs, increased operational efficiency, reduced wastewater flows, reduced costs through industrial symbiosis (sharing management of utilities and services), and increased production capacity due to the increased availability of clean water.[12]

AGRICULTURAL REUSE

Increased competition for water supplies and the increased availability of treated (secondary-treated wastewater) and recycled water (tertiary-treated) provides an opportunity to develop this resource for agricultural production particularly during times of drought when regular water supplies are limited or non-existent. The use of recycled water for irrigated crop production is controlled by regulations that govern the treated water quality, with lesser standards required for forage crops compared with those for food crops. The reuse and recycling of wastewater in agriculture provides numerous benefits including:

- Conserving drinking water supplies
- Easing stress on natural water bodies usually used for extracting irrigation water
- Recycled water containing valuable nutrients reducing fertiliser requirements
- Reducing discharge to the environment.[13,14]

Water Reuse for Urban Agriculture

With rapid urbanisation, many cities are facing increasing water shortages in addition to rising food insecurity. Water reuse in urban agriculture provides a year-round supply of water, providing food, income and employment opportunities to cities, improving the urban landscape, and lessening the pollution load on downstream watercourses.[15]

Case: Fit for Purpose Recycled Water in Queensland

Queensland Urban Utilities' recycled water programme is a fit for purpose programme that offers various qualities, or classes, of recycled water to meet customers' requirements, with the price per kilolitre (kL) decreasing with quality (Table 6.2). Examples of fit for purpose include irrigation of sporting fields or golf courses with controlled public access, washing down of hard surfaces in agriculture industries, irrigation of pastures or fodder crops, irrigation of other non-food crops, irrigation of nurseries or forestry plantations, and below ground irrigation of above ground food crops. Under the programme there are four classes of recycled water:

- *Class A+ recycled water*: This is very highly treated recycled water. This class is usually only required for industrial process water or for the irrigation of minimally processed food crops.

- *Classes A and B recycled water:* This usually requires additional treatment processes above the normal sewage treatment processes that are sufficient to allow effluent to be discharged to the environment.
- *Class C recycled water:* This is the lowest quality of recycled water supplied by Queensland Urban Utilities. This quality is generally produced as a result of the requirements for effluent discharge to the environment, as contained within the sewage treatment plant environmental licenses. This class can often be supplied directly without any additional treatment.[16]

Table 6.2 Recycled water pricing

Class of water recycled	Price
Class A+	By negotiation
Class A	$1.314000 per kL
Class B	$1.216000 per kL
Class C	$0.113000 per kL

POTABLE WATER REUSE

Potable water reuse involves the use of a community's wastewater as a source of drinking water. Two forms of planned potable reuse exist, which are indirect potable reuse (IPR) and Direct potable reuse (DPR).

Indirect Potable Reuse

IPR can be defined as the reclamation and treatment of water from wastewater (often sewage effluent) and the eventual returning of it into the current/natural water cycle well upstream of the drinking water treatment plant. Planned IPR means there is an intent to reuse the water for potable use. The point of return could be either into a major water supply reservoir, a stream feeding a reservoir, or into a water supply aquifer (managed aquifer recharge) where natural processes of filtration and dilution of the water with natural flows aim to reduce real or perceived risks associated with eventual potable reuse. IPR (unplanned) is defined as wastewater entering the natural water (creeks, rivers, lakes, aquifers), which is eventually extracted from the natural system for drinking water: usually with no awareness that the natural system contains wastewater.[17]

Direct Potable Reuse

DPR can be defined as either the injection of recycled water directly into the potable water supply distribution system downstream of the water treatment plant or into the raw water supply immediately upstream of the water treatment plant (injection could be either directly into the water pipeline or into a service reservoir). This means water used by consumers could contain either undiluted or slightly diluted recycled water. The key distinction with IPR is that there is no temporal or spatial separation between the recycled water introduction and its distribution to consumers.[18]

Case: Potable Reuse in the City of Wichita Falls, Texas

The City of Wichita Falls' DPR project went online in 2014 and involves a seven-step process for treating the water. After being processed through the wastewater treatment plant, the treated reuse water is disinfected and pumped into the River Road Resource Recovery Facility to the Cypress Water Treatment Plant. From there the water is treated in the Microfiltration Reverse Osmosis Plant. After which, the reuse water is treated through reverse osmosis. The water is then released into a holding lagoon. The reuse water is then blended with raw lake water on a 50-50 basis. The blended water from the City's water sources lakes, Lake Arrowhead and Lake Kickapoo, is treated through an extensive series of steps to produce safe, clean drinking water ready for distribution. In addition to the DPR project, the city recently completed construction of its IPR project, which returns the highly purified effluent from the River Road Resource Recovery Facility to Lake Arrowhead. The project will return around 8.5 million gallons a day to the reservoir which equates to three billion gallons over a year.[19,20]

Managed Aquifer Recharge

Aquifers are underground rock formations or sedimentary deposits porous enough to hold water. Most aquifers are naturally recharged by rainfall or other surface water that infiltrates into the ground. However, in regions where groundwater use is greater than natural recharge rates, aquifers will be depleted over time. Managed aquifer recharge (MAR) consists of water management methods that recharge an aquifer using either underground recharge techniques or surface water. There are two main approaches used to recharge aquifers: deep injection and surface infiltration. Deep injection methods put excess water directly into the aquifer using wells. Surface infiltration can involve:

- Creating artificial streams and ponds in fast-draining soils
- Creating local catchment systems for rainwater and stormwater
- Diverting water to naturally infiltrating river channels during low-flow seasons.

Existing MAR projects rely mainly on river water, stormwater, and treated wastewater. Treated wastewater provides the most consistent supply and the least competition from other water users. River water is generally only available for aquifer recharge during wetter periods. Stormwater, despite it being sporadic and seasonally available, is a common source for aquifer recharge projects as it decreases flooding and captures water that is otherwise lost as runoff. In practice, many existing MAR projects rely on a combination of all three water sources.[21]

Case: Central Arizona Project Recharge Program

The Central Arizona Project (CAP) Recharge Program is Arizona's single largest resource for renewable water supplies. CAP is designed to bring about 1.5 million acre-feet of water from the Colorado River to Central and Southern Arizona every year. More than five million people, or more than 80% of the state's population, live in Maricopa, Pima, and Pinal counties, where CAP water is delivered. CAP carries water from Lake Havasu to the southern boundary of the San Xavier Indian Reservation southwest

---

of Tucson through a 336-mile long series of aqueducts, tunnels, pumping stations, and pipelines. CAP has developed seven recharge projects and currently operates six. For instance, the Tucson Active Management Area (AMA) recharge facilities currently have a cumulative recharge capacity of 80,000 acre-feet per year and include the Pima Mine Road and Lower Santa Cruz Recharge Projects. In the Phoenix AMA, there are four existing facilities: The Aqua Fria, Hieroglyphic Mountains, Tonopah Desert, and Superstition Mountain recharge projects, with a combined annual permitted capacity of 341,500 acre-feet per annum. The recharge programme allows surface water supplies, such as the Colorado River, to be stored underground now for recovery later during periods of reduced water supply. The benefits of the programme include:

- Encouraging the use of renewable water supplies instead of continued over-reliance on finite groundwater supplies.
- Mitigating of impacts of groundwater overdraft including subsidence and increased energy costs for pumping water from greater depths.
- Eliminating the need to construct costly surface reservoirs that are prone to excessive evaporation losses in Arizona's arid climate.
- Providing an alternative mechanism to deliver CAP water through recharge and recovery instead of constructing costly water treatment plants and distribution facilities.
- Improving the quality of recharged surface water by filtration through underlying sediments in a process known as soil aquifer treatment.[22]

ENVIRONMENTAL REUSE

Environmental reuse is mainly the use of recycled water to support wetlands and to supplement stream and river flows. Wetlands provide many benefits including wildlife habitat, water quality improvement, flood diminishment, and fisheries breeding grounds. For waterways that have been impaired or dried from water diversion, water flow can be augmented with recycled water to sustain and improve the aquatic and wildlife habitat.[23,24]

Case: Sacramento's South County Ag Program

Regional San, in collaboration with regional stakeholders, is developing the South Sacramento County Agriculture and Habitat Lands Recycled Water Program (South County Ag Program). The South County Ag Program will provide multiple benefits including providing a safe and reliable supply of tertiary-treated water for agricultural uses, reducing groundwater pumping, and supporting habitat restoration efforts. The programme proposes to use tertiary-treated recycled water on allowed crops (e.g. alfalfa, irrigated pastures, etc.) and irrigate permanent agricultural lands and habitat mitigation lands through new recycled water transmission and distribution systems. The amount of water saved would be the equivalent to the potable water needs of up to 100,000 homes in the Sacramento region. The programme will help conserve and protect surface water and groundwater supplies by reducing dependence on these resources. Recycled water could help raise groundwater levels enough to help bring increased flows back to the river, with modelling studies indicating that use of recycled water will help recharge the groundwater basin, potentially bringing a 20–30-foot rise in groundwater elevation.[25]

Urban Reuse

In urban areas, onsite non-potable water systems are utilised to meet non-potable needs, such as cooling buildings, irrigating landscapes, and flushing toilets and urinals.[26]

Greywater

Greywater is reusable wastewater from residential, commercial, and industrial bathroom sinks, bathtub shower drains, and clothes washing equipment drains. Greywater is reused onsite, usually for toilet flushing and irrigation. Greywater systems vary significantly in their complexity and size from small systems with simple treatment to large systems with complex treatment processes. Nevertheless, most have common features including a tank for storing the treated water, a pump, a distribution system for transporting the treated water to where it is needed, and some sort of treatment. If greywater systems are used for toilet flushing, it could potentially save a third of the mains water used in the home. Greywater systems also reduce the amount of wastewater generated, transported, and treated at wastewater treatment facilities. Overall, greywater systems can reduce water supply and wastewater bills.[27]

Blackwater

Blackwater, or sewage, is the wastewater from toilets. Blackwater contains urine and excreta, which is combined with flush water and toilet paper and fed into the sewer system. After flushing the blackwater into the municipal sewer system, it is transported to a facility for treatment. The daily amount of blackwater produced is dependent on the type of toilet: the average person generates 1.5 litres of waste per day; however, this amount increases to 25–50 litres per day with the additional quantity of flush water varying with flushing system. In blackwater recycling systems, all the blackwater is routed to an initial tank via gravity, from which it settles, and a primary colony of bacteria eats at the waste. The blackwater then goes through an aeration stage, a sludge settling stage, before it is chlorinated and used as irrigation water (watering lawns or non-food gardens). The benefits of blackwater recycling include:

- *Energy conservation*: The removal of harmful bacteria from blackwater in processing plants is energy-intensive.
- *Water conservation*: Using recycled blackwater to water lawns and non-food gardens helps conserve potable water that would otherwise be used.
- *Resource conservation*: Plants that are grown using recycled blackwater do not need fertiliser as the water is already nutrient-rich, eliminating the need for adding fertilising chemicals.
- *Habitat protection*: Recycling blackwater lessens the chance of wastewater entering water bodies, impacting natural habitats.[28,29]

Rainwater Harvesting

Rainwater harvesting systems collect and store rainfall for later use. When designed appropriately, they slow down and reduce runoff and provide a source of water. There are two main types of rainwater harvesting systems: passive and active. Passive harvesting systems, for example, rain barrels, are typically small volume (50–100 gallon) systems designed to capture rooftop runoff. Rain barrels are usually used in residential applications where the flow from rain gutter downspouts is easily captured for outdoor use, for example, garden and landscape irrigation or car washing. Active harvesting systems, for example, cisterns, are larger volume (typically 1000–100,000-gallon) systems which capture runoff from roofs or other suitable surfaces, provide water quality treatment, and use pumps or sufficient head to supply water to a distribution system. Rainwater collected in active systems is typically used for irrigation or for indoor non-potable water replacement, for example, toilet flushing, clothes washing, evaporative cooling, etc.[30]

Case: New York City's On-Site Water Reuse Grant Pilot Program

New York City's Department of Environmental Protection has implemented the On-Site Water Reuse Grant Pilot Program to provide commercial, mixed-use, and multi-family residential property owners with incentives to install water reuse systems. Grants are available for water reuse systems at the individual building and district level, with district-scale projects involving two or more parcels of land such as a housing development, where the project reduces demand in the shared distribution system. Individual building-scale projects can receive up to $250,000 in reimbursement for a system designed to save at least 32,000 gallons per day (gpd), and district-scale projects are eligible to receive up to $500,000 in reimbursement for a system designed to save at least 94,000 gpd. The NYC Construction Code regulates two types of on-site water reuse systems that can be installed, as follows: (1) Wastewater reuse systems (blackwater, greywater, rainwater) for flushing of toilets and urinals, cooling tower makeup, washing of sidewalks, streets or buildings, laundry, and subsurface or drip irrigation systems and (2) Rainwater reuse systems (rainwater, cooling tower condensate) used solely for cooling tower makeup and subsurface irrigation and drip irrigation.[31]

Stormwater Harvesting

Stormwater harvesting involves collecting, storing, and treating stormwater from urban areas, which can then be used as recycled water. The stormwater is collected from stormwater drains or creeks and the recycled water produced is commonly used to water public parks, gardens, sports fields, and golf courses. The source area of harvested stormwater largely determines the quality of stormwater supply in a stormwater harvesting system: as precipitation accumulates and flows over surfaces it collects pollutants and microbial contaminants. The type of and quantity of pollution in stormwater depends on the composition of the surfaces over which stormwater runoff flows and the activities within the drainage area that generate pollution.

Overall, a stormwater harvesting system usually has four components: A collection system which could include the catchment area and stormwater infrastructure (such as kerbs, gutters, and storm sewers); a storage

Table 6.3 Benefits of stormwater harvesting systems

Benefit	Description
Reduces impacts of urbanisation on watershed hydrology	• Reduces runoff volume from the site • Reduces peak stream flows following storm events • Reduces flooding in downstream waters • Increases groundwater recharge
Reduces impacts of urbanisation on water quality	• Reduces pollutant loads to downstream receiving waters
Increases water conservation	• Conserves potable water for essential uses • Provides an alternative to potable water during time of peak demand • Reduces or limits withdrawals from ground or surface water supply • Maintains reliable water supply in event of municipal service disruption
Reduces stress on existing/need for additional infrastructure	• Reduces the size of stormwater best management practices (BMP) needed to achieve regulatory requirements • Increases the efficiency or extends the life of stormwater BMP/infrastructure • Reduces stress on municipal supply systems during peak usage • Reduces stress on/cost of water supply and treatment infrastructure • Reduces community expenditure on expansion of infrastructure
Energy, education, environment, and economics	• Provides educational opportunities/increases public awareness • Attains sustainable design certification/recognition • Reduces consumption of potable water for cost savings • Reduces the energy footprint of water, wastewater, and stormwater infrastructure • Reduces onsite erosion and flooding

unit such as a cistern or pond; a treatment system, pre and post, that removes solids, pollutants, and microorganisms; and a distribution system such as pumps, pipes, and control system. The potential benefits of stormwater harvesting are listed in Table 6.3.[32]

Case: Los Angeles' Industrial Stormwater Rebate Program

The Los Angeles Department of Water and Power (LADWP) is launching the Industrial Stormwater Rebate Program (ISRP) for the infiltration and on-site use of stormwater from industrial facilities. The programme aligns with the city's goals of reducing the purchase of imported water by 50% by 2025 and producing 50% of Los Angeles' water locally by 2035. Under the proposed programme, all industrial facilities regulated by the Industry General Permit and within LADWP's service area/areas of interest will be eligible to participate in the ISRP. The total rebate amounts will be based on yield for each respective facility, with the rebate amount of $1100/acre foot (AF) for infiltration and $1550/AF for onsite reuse. Requirements for the programme may include programme participation agreement, possible flow meter installation, subject to inspection and verification, and access to the stormwater infrastructure for testing, inspection, and observation regarding maintenance, operation, repair, and replacement of the stormwater infrastructure. The programme has an initial funding of up to $800,000, which is projected to cover up to 12 applicants. Based on the first year, the programme could result in the capture of up to 4000 AF per year of stormwater.[33]

NOTES

1. Nikolaos Voulvoulis, "Water Reuse from a Circular Economy Perspective and Potential Risks from an Unregulated Approach," *Current Opinion in Environmental Science & Health* 2 (2018).
2. US EPA, "Water Reuse and Recycling: Community and Environmental Benefits," https://www3.epa.gov/region9/water/recycling/#benefits.
3. C. Schaum, D. Lensch, and P. Cornel, "Water Reuse and Reclamation: A Contribution to Energy Efficiency in the Water Cycle," *Journal of Water Reuse and Desalination* 5, no. 2 (2014).

4. EPA Victoria, "Guidelines for Environmental Management Use of Reclaimed Water," (2003), https://www.epa.vic.gov.au/~/media/Publications/464%202.pdf.
5. "Reusing and Recycling Water," https://www.epa.vic.gov.au/your-environment/water/reusing-and-recycling-water#sewage.
6. Australian Water Association, "Water Recycling," http://www.awa.asn.au/AWA_MBRR/Publications/Fact_Sheets/Water_Recycling_Fact_Sheet/AWA_MBRR/Publications/Fact_Sheets/Water_Recycling_Fact_Sheet.aspx?hkey=54c6e74b-0985-4d34-8422-fc3f7523aa1d.
7. A. Pain and D. Spuhler, "Wastewater Reuse in Industry," https://sswm.info/water-nutrient-cycle/water-use/hardwares/optimisation-water-use-industries/wastewater-reuse-in-industry.
8. Ibid.
9. Centre for Advanced Process Technology for Urban REsource recovery, "Water 'Fit for Use'," https://capture-resources.be/water-fit-use.
10. Spuhler, "Wastewater Reuse in Industry."
11. Evides Industriewater, "Process Water," https://www.evidesindustriewater.nl/products-markets/process-water/?lang=en.
12. Evoqua, "Industrial Water Reuse and Recycle," http://www.evoqua.com/en/brands/IPS/Pages/Industrial-Water-Recycle.aspx.
13. University of California Agriculture and Natural Resources, "Use of Treated Wastewater for Crop Production," (2017), https://anrcatalog.ucanr.edu/pdf/8534.pdf.
14. Hunter Water, "Agricultural Use," https://www.hunterwater.com.au/Water-and-Sewer/Recycling--Reuse/Agricultural-Use.aspx.
15. R. C. Brears, *The Green Economy and the Water-Energy-Food Nexus* (London, UK: Palgrave Macmillan, 2017).
16. Queensland Urban Utilities, "Recycled Water," https://www.urbanutilities.com.au/business/business-services/recycled-water.
17. Recycled Water in Australia, "Definitions," http://www.recycledwater.com.au/index.php?id=105.
18. Ibid.
19. City of Wichita Falls, "What Is the Latest News on the Direct Potable Reuse Project (DPR)?" http://www.wichitafallstx.gov/faq.aspx?qid=513.
20. "2018 Drinking Water Quality Report," (2018), http://www.wichitafallstx.gov/DocumentCenter/View/30928/2018-Drinking-Water-Report-for-Web.
21. American Geosciences Institute, "Managed Aquifer Recharge," https://www.americangeosciences.org/critical-issues/factsheet/managed-aquifer-recharge.
22. Central Arizona Project, "About Us," https://www.cap-az.com/about-us; "Recharge Program," https://www.cap-az.com/departments/recharge-program.

23. US EPA, "Water Reuse and Recycling: Community and Environmental Benefits."
24. Justin E. Lawrence et al., "Recycled Water for Augmenting Urban Streams in Mediterranean-Climate Regions: A Potential Approach for Riparian Ecosystem Enhancement," *Hydrological Sciences Journal* 59, no. 3–4 (2014).
25. Regional San, "South County Ag Program," https://www.regional-san.com/south-county-ag-program; "Water Recycling Program South County Ag Fact Sheet," (2014), https://www.regionalsan.com/sites/main/files/file-attachments/final_benefit_only_tm.pdf.
26. US Water Alliance, "Making the Utility Case for Onsite Non-potable Water Systems," (2018), http://uswateralliance.org/sites/uswater-alliance.org/files/publications/NBRC_Utility%20Case%20for%20 ONWS_032818.pdf.pdf.
27. Environment Agency, "Greywater for Domestic Users: An Information Guide," (2011).
28. Hamburg Wasser, "Blackwater," https://www.hamburgwatercycle.de/en/hamburg-water-cycler/blackwater/.
29. Worcester Polytechnic Institute, "Blackwater Recycling System," https://wp.wpi.edu/capetown/projects/p2009/gardens/accomplishments/composting-options-for-the-proposed-sanitation-facility/black-water-recycling-system/.
30. US EPA, "Rainwater Harvesting. Conservation, Credit, Codes, and Cost Literature Review and Case Studies," (2013), https://www.epa.gov/sites/production/files/2015-11/documents/rainharvesting.pdf.
31. R. C. Brears, January 9th, 2018, https://medium.com/mark-and-focus/new-york-city-embracing-future-water-challenges-a4ced29bc77f.
32. Minnesota Pollution Control Agency, "Overview for Stormwater and Rainwater Harvest and Use/Reuse," https://stormwater.pca.state.mn.us/index.php?title=Overview_for_stormwater_and_rainwater_harvest_and_use/reuse.
33. LADWP, "Industrial Stormwater Rebate Program," (2018), https.//www.ladwp.com/cs/idcplg?IdcService=GET_FILE&dDocName=OPLADWPCCB645433&RevisionSelectionMethod=LatestReleased.

References

American Geosciences Institute. "Managed Aquifer Recharge." https://www.americangeosciences.org/critical-issues/factsheet/managed-aquifer-recharge.
Australian Water Association. "Water Recycling." http://www.awa.asn.au/AWA_MBRR/Publications/Fact_Sheets/Water_Recycling_Fact_Sheet/AWA_MBRR/Publications/Fact_Sheets/Water_Recycling_Fact_Sheet.aspx?hkey=54c6e74b-0985-4d34-8422-fc3f7523aa1d.

Brears, R. C. *The Green Economy and the Water-Energy-Food Nexus*. London, UK: Palgrave Macmillan, 2017.
———. "New York City Embracing Future Water Challenges." Mark and Focus, 2018.
Central Arizona Project. "About Us." https://www.cap-az.com/about-us.
———. "Recharge Program." https://www.cap-az.com/departments/recharge-program.
Centre for Advanced Process Technology for Urban REsource recovery. "Water 'Fit for Use'." https://capture-resources.be/water-fit-use.
City of Wichita Falls. "2018 Drinking Water Quality Report." (2018). http://www.wichitafallstx.gov/DocumentCenter/View/30928/2018-Drinking-Water-Report-for-Web.
———. "What Is the Latest News on the Direct Potable Reuse Project (DPR)?" http://www.wichitafallstx.gov/faq.aspx?qid=513.
Environment Agency. "Greywater for Domestic Users: An Information Guide." (2011).
EPA Victoria. "Guidelines for Environmental Management Use of Reclaimed Water." (2003). https://www.epa.vic.gov.au/~/media/Publications/464%202.pdf.
———. "Reusing and Recycling Water." https://www.epa.vic.gov.au/your-environment/water/reusing-and-recycling-water#sewage.
Evides Industriewater. "Process Water." https://www.evidesindustriewater.nl/products-markets/process-water/?lang=en.
Evoqua. "Industrial Water Reuse and Recycle." http://www.evoqua.com/en/brands/IPS/Pages/Industrial-Water-Recycle.aspx.
Hamburg Wasser. "Blackwater." https://www.hamburgwatercycle.de/en/hamburg-water-cycler/blackwater/.
Hunter Water. "Agricultural Use." https://www.hunterwater.com.au/Water-and-Sewer/Recycling–Reuse/Agricultural-Use.aspx.
LADWP. "Industrial Stormwater Rebate Program." (2018). https://www.ladwp.com/cs/idcplg?IdcService=GET_FILE&dDocName=OPLADWPCCB645433&RevisionSelectionMethod=LatestReleased.
Lawrence, Justin E., Christopher P. W. Pavia, Sereyvicheth Kaing, Heather N. Bischel, Richard G. Luthy, and Vincent H. Resh. "Recycled Water for Augmenting Urban Streams in Mediterranean-Climate Regions: A Potential Approach for Riparian Ecosystem Enhancement." *Hydrological Sciences Journal* 59, no. 3–4 (2014, April 3): 488–501.
Minnesota Pollution Control Agency. "Overview for Stormwater and Rainwater Harvest and Use/Reuse." https://stormwater.pca.state.mn.us/index.php?title=Overview_for_stormwater_and_rainwater_harvest_and_use/reuse.
Pain, A., and D. Spuhler. "Wastewater Reuse in Industry." https://sswm.info/water-nutrient-cycle/water-use/hardwares/optimisation-water-use-industries/wastewater-reuse-in-industry.

Queensland Urban Utilities. "Recycled Water." https://www.urbanutilities.com.au/business/business-services/recycled-water.

Recycled Water in Australia. "Definitions." http://www.recycledwater.com.au/index.php?id=105.

Regional San. "South County Ag Program." https://www.regionalsan.com/south-county-ag-program.

———. "Water Recycling Program South County Ag Fact Sheet." (2014). https://www.regionalsan.com/sites/main/files/file-attachments/final_benefit_only_tm.pdf.

Schaum, C., D. Lensch, and P. Cornel. "Water Reuse and Reclamation: A Contribution to Energy Efficiency in the Water Cycle." *Journal of Water Reuse and Desalination* 5, no. 2 (2014): 83–94.

University of California Agriculture and Natural Resources. "Use of Treated Wastewater for Crop Production." (2017). https://anrcatalog.ucanr.edu/pdf/8534.pdf.

US EPA. "Rainwater Harvesting. Conservation, Credit, Codes, and Cost Literature Review and Case Studies." (2013). https://www.epa.gov/sites/production/files/2015-11/documents/rainharvesting.pdf.

———. "Water Reuse and Recycling: Community and Environmental Benefits." https://www3.epa.gov/region9/water/recycling/#benefits.

US Water Alliance. "Making the Utility Case for Onsite Non-potable Water Systems." (2018). http://uswateralliance.org/sites/uswateralliance.org/files/publications/NBRC_Utility%20Case%20for%20ONWS_032818.pdf.pdf.

Voulvoulis, Nikolaos. "Water Reuse from a Circular Economy Perspective and Potential Risks from an Unregulated Approach." *Current Opinion in Environmental Science & Health* 2 (2018, April 1): 32–45.

Worcester Polytechnic Institute. "Blackwater Recycling System." https://wp.wpi.edu/capetown/projects/p2009/gardens/accomplishments/composting-options-for-the-proposed-sanitation-facility/black-water-recycling-system/.

CHAPTER 7

Developing the Circular Water Economy: Recover

Abstract Wastewater treatment plants in the circular water economy provide multiple opportunities including generating renewable energy and recovering nutrients. In addition, the facilities and infrastructure of water and wastewater treatment facilities provide numerous opportunities to generate renewable energy.

Keywords Water treatment · Resource recovery · Renewable energy

INTRODUCTION

In the circular water economy, wastewater treatment plants are not waste disposal facilities, rather they are resource recovery facilities that in addition to producing clean water, generate renewable energy, and recover nutrients. In addition, the facilities and infrastructure of water and wastewater treatment facilities provide opportunities to generate renewable energy. This chapter will first discuss renewable energy generation derived from wastewater treatment processes, before providing an overview of traditional renewable energy activities at water and wastewater treatment facilities. Finally, the chapter will discuss the recovery of resources from wastewater in the development of the circular water economy.

© The Author(s) 2020
R. C. Brears, *Developing the Circular Water Economy*,
Palgrave Studies in Climate Resilient Societies,
https://doi.org/10.1007/978-3-030-32575-6_7

RENEWABLE ENERGY GENERATION TECHNOLOGIES AT WASTEWATER TREATMENT FACILITIES

Energy derived from wastewater treatment is a renewable energy resource and can include electrical energy, heat or biofuels from utilisation of digester gas (biogas that consists mainly of methane and carbon dioxide), electrical energy and heat from thermal conversion of biomass (biosolids), electrical energy from biosolids products used by other entities (e.g. pellets used in power plants or industrial furnaces), and heating or cooling energy using plant influent or effluent as a heat source or sink for a heat pump.[1,2]

Biogas from Anaerobic Digestion

Anaerobic digestion is a series of biological processes in which microorganisms break down biodegradable material in the absence of oxygen. One of the end products is biogas, which is combusted to generate electricity and heat or can be processed into renewable natural gas and transportation fuels. Wastewater treatment plants employ anaerobic digesters to break down sewage sludge and eliminate pathogens in wastewater. Captured biogas is transported via pipe in the digester either directly to a gas use device or to a gas treatment system (for e.g. moisture or hydrogen sulphide removal). Captured biogas can also be upgraded by removing carbon dioxide, nitrogen, and oxygen in order to meet high purity and British Thermal Units (BTU) requirements for pipeline injection or compressed biomethane vehicle fuel.[3]

Case: Hamburg Wasser's Sewage Gas Becomes Urban Biogas

In Hamburg, Germany, Hamburg Wasser's 10 digester towers, with a total volume of 80,000 cubic metres, contains sewage sludge that rots continuously at a constant temperature of around 36 degrees Celsius. Under anaerobic conditions, 95,000 cubic metres of digester gas is produced per day. This gas is either converted directly via a generator into electricity on site and used or is fed as biomethane into the urban gas network. Urban customers can choose the wastewater treatment plant-associated urban biogas content in their gas supply with two different pricing options available: Alster Shore: The cheap entry-level tariff has at least 1%

urban biogas content with the tariff charged dependent on the postal code of the customer and their annual consumption (a kWh estimate) and *Alster Pearl*: The premium gas tariff has at least 5% urban biogas content with the tariff charged dependent on the postal code of the customer and their annual consumption (a kWh estimate).[4]

Case: Portland's Wastewater Treatment Plant's Renewable Natural Gas Project

Portland's Columbia Boulevard Wastewater Treatment Plant Renewable Natural Gas Facility Project will convert biogas (which contains 60% methane) into renewable natural gas (RNG). The biogas (methane), which forms in the plant's digesters, will be sent for processing onsite. The RNG produced will then be distributed through an NW Natural pipeline to a new on-site fuelling station built for City vehicles. The project will cut greenhouse gas emissions by 21,000 tonnes per annum, generate over $3 million in revenue per year for the City, and replace 1.34 million gallons of diesel fuel with RNG, enough to power 154 garbage trucks a year. Currently, the wastewater treatment plant uses around half of the plant's waste methane to heat and power the treatment plant while a quarter of the waste methane is flared. Overall, the project will move the plant to 100% methane recovery.[5]

Co-digestion

Anaerobic digestion of energy-rich organic waste materials including restaurant grease and food waste along with wastewater treatment sludge is defined as 'co-digestion'. In addition to diverting food waste and fats, oils, and grease (FOG) away from landfills and collection systems, co-digestion provides multiple benefits including greenhouse gas mitigation (avoids the release of methane from landfills) and economic savings (cost-recovery from producing on-site power).

Case: The Municipality of Anyang's Wastewater Treatment Plant Co-digesting Food Waste

A new underground wastewater treatment plant serving 700,000 people in the Municipality of Anyang, South Korea, will use thermal hydrolysis to co-digest organic waste. The plant will co-digest about 27,000 dry tonnes of organic waste per year, of which 65% is sewage sludge and the remaining 35% food waste. Biogas produced from the co-digestion plant will be turned into electricity and heat for the processes. The remaining high dry solids dewatered product after digestion will be dried and blended with millet grass to produce biomass fuel for co-firing in existing power plants. In addition to the operational and environmental benefits, the plant is located underground, allowing the Municipality of Anyang to benefit from a small footprint and substantial savings in capital expenditures associated with the construction of the digesters.[6]

Biogas for Combined Heat and Power

Combined heat and power (CHP) are the most prevalent means of utilising biogas. As the process of anaerobic digestion requires some heat it is suited to CHP. The ratio of heat to power varies depending on the scale and technology but typically 35–40% is converted to electricity, 40–45% to heat and the balance lost as inefficiencies in the various stages of the process. This equates to over 2kWh electricity and 2.5kWh heat per cubic metre, at 60% methane.[7] CHP offers a variety of benefits including:

- *Efficiency:* CHP requires less fuel than separate heat and power generation to produce a given energy output. CHP also avoids transmission and distribution losses that occur when electricity travels over power lines from central generating units.
- *Reliability:* CHP can provide high-quality electricity and thermal energy to a site regardless of what happens on the power grid, decreasing the impacts of outages and improving power quality for sensitive equipment.

- *Environmental*: Because less fuel is burned to produce each unit of energy output, CHP reduces greenhouse gases and other air pollutants.
- *Economic*: CHP lowers a facility's energy bill considerably due to its high efficiency and it can provide a hedge against unstable energy costs.[8]

Case: City of Charlotte's CHP Facility at McAlpine Creek Wastewater Treatment Plant

The City of Charlotte's McAlpine Creek Wastewater Treatment Plant produces on an average day 900 pounds of methane gas which is burned off as a greenhouse gas and waste product. To harness the potential of this gas by turning it into a source of renewable energy the city opened the CHP facility at the plant. The facility is one of the largest producers of renewable energy in the Charlotte area and the only facility of its kind at a wastewater treatment plant in North Carolina. The CHP facility burns methane gas in a specialised engine that runs a generator and produces power and heat from the engine. The heat is recycled back to the anaerobic digestion system while the power generated by the CHP is sold back to Duke Energy, offsetting the plant's energy costs. The CHP generator creates around 656,000 kWh of energy each month and to date, more than 5.3 million kWh of energy has been produced around-the-clock at the facility.[9]

Thermal Conversion of Wastewater and Biosolids

The process of converting biosolids to energy is either through anaerobic biodegradation (as discussed above) or through thermal conversion. Thermal oxidation (incineration), which is the complete oxidation of organics (biomass) to carbon dioxide and water in the presence of excess air, is a well-established technology. The benefits of thermal conversion include reduction in biosolids mass, generation of heat for use in heating or electricity generation, reduction in the facility's overall carbon footprint, lowering the reliance on fossil fuels, generation of ash for use in building materials, and generation of additional revenue to utilities.[10]

Case: Hong Kong's Futuristic Waste-to-Energy Facility

Each day Hong Kong produces nearly three million cubic metres of sewage which equates to over 1200 tonnes of sludge. Rather than relying on landfill as the only means of sludge disposal, Hong Kong has launched T · PARK, which is a futuristic waste-to-energy facility. Over the years, landfilling has been the solution to disposing of sludge, however, this is not a sustainable solution due to a substantial increase in the amount of sludge requiring disposal as a result of population growth over the past decades and the upgrading and improvement of the sewage treatment systems, including the Harbour Area Treatment Scheme, initiated in 2015, which coordinates the collection of both sides of Victoria Harbour. As a result, it is estimated that the amount of sludge will grow to 2000 tonnes per day in 2030. T · PARK's solution is the burning of the sewage sludge through advanced incineration technology. The plant has the capacity to handle a maximum capacity of 2000 tonnes of sludge per day with the heat generated recovered and turned into electricity that can meet the demand of the entire facility. When running at full capacity the plant can produce up to 2 MW of surplus electricity, enough to support 4000 households. After incineration, the sludge is converted into ash and the residues—90% less of the total original sludge volume—are disposed of in landfill, which reduces emissions of greenhouse gases by up to 237,000 tonnes per year.[11]

Thermal Energy Recovery from Wastewater

Thermal energy can be recovered from raw wastewater or effluent by exploiting the significant temperature differential between wastewater and ambient conditions. This temperature difference can be recovered for use in heating and cooling systems, which is often used for buildings at the facility and in buildings of areas surrounding the facility.[12]

Case: Heat from Wastewater in Berlin

In Berlin, Germany, IKEA Berlin-Lichtenberg is supplied with energy via a special wastewater pressure line from Berliner

Wasserbetriebe (500,000–1.4 million litres of wastewater flows through the line per hour). The wastewater, with a temperature between 12 and 15 degrees Celsius heats and cools the store 365 days a year. During winter, special heat pumps withdraw the heat from the wastewater, heating the water to around 35 degrees Celsius and then channelling it to the floor heating and radiant ceiling panels. During the summer months, the heat of the store is drained into the wastewater system. This type of energy use allows the store to cover 70% of its energy demand in winter and 100% in the summer. During the cold season, the system is supported by a gas boiler.[13]

Traditional Renewable Energy at Water and Wastewater Treatment Facilities

Water utilities can implement traditional renewable energy activities on facility-owned buildings and surrounding land, including the following.

Solar Radiation Captured at Facilities

Solar photovoltaic (PV) systems directly convert sunlight into electricity using solar cells. These systems, which can produce electricity even in the absence of strong sunlight, can generate significant quantities of electricity depending on a variety of factors including quality of the sunlight and the system's mounted pitch. PV systems can be installed on rooftops, making them ideal for areas where open space is limited.[14] Solar PVs can be a cost-effective option for water utilities with solar radiation captured at both water and wastewater treatment facilities.

Case: Australia's First Solar Array on a Water Treatment Plant

Wannon Water, in Victoria, Australia, has installed a solar system on its Hamilton Water Treatment Plant, the first large-scale solar system installed on a roof of an Australian water utility tank. A total of 334 high-efficiency solar PV panels are located on the roof of the clear water storage tank, reducing the plant's demand on the electricity grid by 25% and reducing greenhouse gas emissions by 150,000 kilograms per annum. This new system is part of the utility's pledge to achieve net-zero emissions by 2050, with an interim target of 40% reduction by 2025. The system, which came at a net cost of $125,000, has a payback period of seven years through reduced energy usage. It has an estimated operating life of 25 years. At peak performance, the entire water treatment plant, which provides drinking water for Hamilton, Dunkeld, and Tarrington, can be 100% powered by renewable energy, with any excess being exported to neighbouring properties via a grid connection. The average renewable annual energy output is more than 125,000-kilowatt-hours, which is the equivalent to providing clean energy for 22 average residential homes.[15]

Floating Solar Photovoltaic Systems

The use of land PV systems can be partially or completely avoided by implementing floating PV systems (FPVSs). FPVSs are usually installed on enclosed water bodies such as reservoirs, ponds, and small lakes. Due to their novelty, most systems are proprietary and of small-medium size. There are numerous technical benefits from installing FPVSs, including the following:

- The evaporative cooling of PV modules and cables caused by the water body increases the efficiency of the system.
- They reduce evaporation of the free surface of the water, conserving the volume of stored water.
- They reduce algae growth.
- FPVSs reduce the formation of waves and therefore limit the erosion of the banks of a reservoir.
- FPVSs do not require a land area, providing an economic advantage.
- The reflectivity (albedo) of the water increases the incidence of radiation in the PV array and, therefore, its energy generation.[16]

Case: United Utilities' Floating Solar Farm

United Utilities in the United Kingdom is building a floating solar farm on the surface of Langthwaite Reservoir. The power generated will be used to run the neighbouring Lancaster water treatment works which supplies water to 152,000 people across Lancaster, Morecambe, and Heysham. The floating array will be around 7200 square metres in size with just over 3500 solar panels. The installation will cover the size of a football pitch and will provide 1 MW of power: the equivalent of the needs of 200 homes. This will be the utility's second floating solar installation with the company having installed Europe's first commercial floating solar array at its Godley reservoir near Manchester in 2016. That array is three times the size of the one at Lancaster and can generate 3 GWh of electricity per year.[17]

Wind Power Captured at Facilities

Wind energy, which is captured on site using wind turbines, can be very cost-effective in areas with adequate wind resources. As opposed to large utility-scale wind farm turbines, which can have capacities as high as 3 MW, small wind turbines are often better suited for local facilities. Wind turbines are most often installed in non-urban areas because installations typically require at least one acre of land and wind speeds averaging around 24 kilometres per hour at 50 metres above the ground.[18]

Case: Berliner Wasserbetriebe's Wind Turbines

In Berlin, Germany, treating wastewater requires as much energy as a city of 280,000 inhabitants. To reduce energy demand and carbon emissions, Berliner Wasserbetriebe has installed three wind turbines next to the Schönerlinde sewage treatment plant with a total capacity of 2 MW. For the wastewater treatment plant, this is a significant step towards energy self-sufficiency as the share of self-generated electricity has increased from around 30% in the past to 80% now. Prior to the installation of the wind turbines energy in the treatment plant was generated primarily through the power generation of the sewage gas in a CHP plant with a micro gas turbine. The wind turbines, each with a rotor diameter of almost 93 metres, enables the utility to avoid up to 7000 tonnes of carbon dioxide emissions each year.[19]

Energy Recovery Hydropower

Energy recovery hydropower can be defined as "*hydropower built using an existing, pressurized, manmade water conveyance that is already diverting water from a natural waterway for the distribution of water for agricultural, municipal, or industrial consumption and not primarily for the generation of electricity*". Energy recovery from hydropower differs from some forms of conventional hydropower in that it is not located on natural rivers or waterways.[20]

Case: Athens Water and Sewerage Company Maximising Its Hydropower Potential

Athens Water Supply and Sewerage Company (EYDAP) in Greece has launched an ambitious programme of renewable energy utilisation with the objective of contributing to the optimisation of the energy balance across the country and society and exploring the possibility of expanding to new profitable business. Along the aqueduct system that brings raw water into the Athens area for treatment, there are existing energy dissipation works (small waterfalls along the aqueduct). EYDAP is in the process of converting this energy dissipation works into energy production plants, therefore taking advantage of the aqueducts' hydropower potential. These energy production plants consist of temporarily redirected water flowing through a turbine that converts the hydraulic energy potential into electricity by use of a generator. The water then re-joins the aqueduct at a lower elevation and continues its flow to the water treatment plant. To date, EYDAP has constructed six small hydropower plants along its aqueducts located in Kirfi (760 kW), Elikona (650 kW), Kartala Kihairona (1200 kW), Mandra (630 kW), Evinos Dam (820 kW), and Klidi (590 kW).[21]

Resource Recovery from Wastewater

In the circular water economy, numerous resources can be recovered from wastewater.

Nitrogen and Phosphorous Recovery from Wastewater

The use of nitrogen and phosphorous-based synthetic fertilisers shows an increasing trend, however, this has led to a large-scale influx of reactive nitrogen into the environment, impacting human health and the environment. Meanwhile, phosphorous, which is a non-renewable resource, is at risk of depletion. Therefore, the recovery and reuse of nitrogen and phosphorous are highly necessary.[22]

Nitrogen

Nitrogenous materials present in the sewage can be removed from sewage effluent and converted into biomass through activated secondary treatment processes. Fertiliser grade ammonium sulphate can be produced from the high ammonia-nitrogen concentration sidestreams from sludge digestion processes by stripping and absorption. The stripping of ammonia can be done by steam (steam is blown through the water and after condensation, a concentrated ammonia solution is produced) or air (air is bubbled through wastewater and takes up the gaseous ammonia). Zeolites and other minerals like clay can be used to absorb ammonium.[23,24]

Case: Vestfjorden Wastewater Treatment Plant in Oslo Recovering Nitrogen

The Vestfjorden Wastewater Treatment Plant in Oslo, Norway is cooperating with Yara, a mineral fertilisers and industrial/environmental solutions company, to recover and reuse nitrogen. Around 12–15% of the total nitrogen load entering the plant is recovered—after anaerobic digestion of sludge, lime conditioning, and filter pressing stages—via ammonia stripping and subsequent capturing (scrubbing) of the ammonia gas with a concentrated nitric acid solution. This stripping and scrubbing treatment results in an industrially usable ammonium nitrate sidestream. As part of this

---

circular business model, Yara supplies the nitric acid and receives the ammonium nitrate sidestream while the wastewater treatment plant benefits from a safe, efficient solution that lowers its operational costs, enabling it to focus on its core business (water cleaning).[25]

Phosphorous

There are two main possibilities of recovering phosphorous from municipal used water: recovery from used water treatment and recovery from produced sludge. Recovery from sewage sludge results in, for example, magnesium ammonium phosphate (MAP), calcium phosphate, and iron phosphate. MAP is commonly referred to as struvite and can easily be separated from used water due to its specific gravity. Phosphorous can be recovered from sludge through supercritical water oxidation, which involves the destruction of organics in the sewage sludge to leave a slurry of inorganic ash in a pure water phase free from organic contaminants. Components, such as phosphorous and coagulants, can be easily recovered from the residual ash. Most processes involved in the recovery of a phosphorous product require chemical consumption. Crystallisation is one established technology with the highest percentage of recovered resource for phosphorous, with a recovery rate of over 90%.[26]

Case: City of Saskatoon's Nutrient Recovery Facility

The City of Saskatoon's Nutrient Recovery Facility at the Wastewater Treatment Plant is the first Canadian site to feature the patented Ostara technology that eliminates the build-up of cement-like material (struvite) in the wastewater treatment equipment. The elimination of this material forming inside the equipment and pipes throughout the plant will prevent significant recurring maintenance and repair costs. The Nutrient Recovery Facility will recover phosphorous and nitrogen from the wastewater and turn it into a highly pure, slow release, environmentally friendly fertiliser called Crystal Green®. The city will produce 240–350 tonnes of fertiliser annually, which will be distributed by Ostara, creating an annual revenue stream for the city. Crystal Green® is an all-in-one fertiliser that contains magnesium, phosphorous, and nitrogen and is slow release which reduces leaching and runoff. It will be able to be blended with other fertiliser products and soil types for application on turf grass, golf courses, and high-value crops including berries and fruit trees.[27]

Land Application of Composting of Biosolids

Wastewater treatment plants produce liquid effluent that is discharged into water bodies or reused as well as a by-product of solid residues (sewage sludge) that must be managed in an environmentally responsible manner. Although the terms 'biosolids' and 'sewage sludge' are often used interchangeably, they are not the same. With further treatment, sewage sludge can yield biosolids, which is defined by the US EPA as "*nutrient-rich organic materials resulting from the treatment of domestic sewage in a treatment facility… that can be recycled and applied as fertilizer to improve and maintain productive soils and stimulate plant growth*". Biosolids that are land applied have been treated to minimise odours and to reduce or eliminate pathogens. The benefits of biosolids for both soil and vegetation are numerous. For instance, biosolids provide primary nutrients (nitrogen and phosphorous) and secondary nutrients such as calcium, iron, magnesium, and zinc. The use of biosolids increases crop yields and maintains nutrients in the root zone and unlike chemical fertilisers, biosolids provide nitrogen that is slowly released over the growing season as the nutrient is mineralised and made available for plant uptake.[28]

Case: Certified Sludge for Agricultural Use in Gothenburg, Sweden

At Ryaverket Wastewater Treatment Plant in Gothenburg, Sweden, sludge goes into the biogas plant where it lies in a 37-degree oxygen-free environment for around three weeks and is digested. After digestion, the sludge is centrifuged to remove additional water. Afterwards, the sludge contains around 70% water and looks like moist soil, part of which is composted and used as plant soil. More than a third of the sludge is sanitised and used as a fertiliser. The sludge is certified according to REVAQ, which is operated by the

Swedish Water and Wastewater Association, the Federation of Swedish Farmers, the Swedish Food Federation, and the Swedish food retailer's federation, in close cooperation with the Swedish Environmental Protection Agency. The goals of REVAQ is to avoid the unacceptable accumulation of metals or undesired organic substances on agricultural land in the long term, have no accumulation of cadmium taking place from 2025, and reduce the accumulation of non-essential substances to a maximum of 0.2% from 2025.[29,30]

Plastic

One of the most non-traditional technologies under development is the production of biodegradable plastic using polymers isolated from biosolids. Polymers contain carbon, hydrogen, oxygen, and nitrogen and therefore biological wastewater can be used to make polymers. Polymers called polyhydroxyalkanoates (PHA) can be produced by anaerobic bacteria by metabolising renewable organic carbon sources. PHA polymers are biodegradable thermoplastics and can be used as a substitute for conventional petroleum-based plastics.[31]

Case: The World's First Kilogram of PHA Produced from Bacteria

In 2015, three Dutch water authorities, Brabantse Delta, De Dommel, and Wetterskip Fryslân, in collaboration with STOWA (Dutch Foundation for Applied Water Research), sludge treatment plant SNB, and two commercial parties, Veolia and KNN, produced the world's first kilogram of PHA produced by bacteria from a wastewater treatment facility in the Dutch province of Zeeland. While the capacity of the project is small—a few kilograms a week—the aim is to scale this up to include the total treated wastewater volume and ultimately result in production capacity of 2000 metric tonnes/year.[32]

Bricks

Sewage sludge ash is the by-product produced during the combustion of dewatered sewage sludge in an incinerator. Sewage sludge ash is primarily a silty material with some sand-sized particles. The size range and properties of the sludge ash depend on the type of incineration system and the chemical additives used in the wastewater treatment process. In the circular water economy, ash from sewage sludge incineration can be used in the brick and tile industry.[33,34,35]

Case: Bricks from Recycled Sewage Waste

Thames Water has signed a deal with a private sector contractor to create energy-efficient bricks from human waste. Each day, wastewater enters Europe's largest sewage works in Beckton with the leftover solids used in the utility's waste-to-energy incinerator. Until now the left-over ash was disposed of in landfill. Thames Water will now provide the contractor with the dried residue ash needed to create the bricks with the ash to be reacted and mixed with carbon dioxide, water, sand, and a small quantity of cement to form aggregate for 17-kilogram blocks. The process captures more carbon dioxide than the initial incineration creates, meaning the bricks act as small-scale carbon storage. Under the terms of the deal, Thames Water will supply ash to make 18,000 tonnes of the aggregate, enough to create 2.3 million heavy-duty bricks.[36]

Mining Wastewater for Metals

Metals can be potentially mined from wastewater, for instance, silver and cadmium are increasingly being found in wastewater and are expensive enough to potentially warrant recovery.[37]

Case: Demineralized Water Plant Trial

The Demineralized Water Plant (DWP) in the Botlek area of the Netherlands owned by Evides is a large-scale demonstration of the

ZERO BRINE project that is coordinated by TU Delft that aims to prove that minerals, such as magnesium, and clean water, can be recuperated from industrial processes for reuse in other industries. The DWP consists of a combination of ion exchanges and membrane technology: dissolved air flotation, reverse osmosis, and mixed bed ion exchange. Waste heat and wastewater streams will be combined in a multi-company site environment eliminating brine effluent (target: zero liquid discharge) of the industrial water supplier, recovering high purity magnesium products (target: magnesium purity>90%), NaCl solution and sulphate salts and recycling streams within the site (target:>70% internal recycling of materials recovered).[38]

NOTES

1. Water Environment Federation, "Water Environment Federation Position Statement: Renewable Energy Generation from Wastewater," (2011), https://www.wef.org/globalassets/assets-wef/5---advocacy/policy-statements/policy-statements/wef-positionstatementonrenewable-energybotfinal14oct2011-1.pdf.
2. R. C. Brears, *The Green Economy and the Water-Energy-Food Nexus* (London, UK: Palgrave Macmillan, 2017).
3. American Biogas Council, "What Is Anaerobic Digestion?," https://www.americanbiogascouncil.org/biogas_what.asp.
4. Hamburg Wasser, "The Sewage Treatment Plant Hamburg," https://www.hamburgwasser.de/privatkunden/unser-wasser/der-weg-des-wassers/abwasserreinigung/klaerwerk-hamburg/.
5. City of Portland, "Columbia Boulevard Wastewater Treatment Plant Renewable Natural Gas Facility Project Pub Meeting," (2017), https://www.portlandoregon.gov/cbo/article/645633.
6. Waste Management World, "Food Waste to Be Co-digested into Biogas in Korean Underground Wastewater Treatment Plant," https://waste-management-world.com/a/food-waste-to-be-co-digested-into-biogas-in-korean-underground-wastewater-treatment-plant.
7. NNFCC Biocentre, "Biogas," http://www.biogas-info.co.uk/about/biogas/.
8. US EPA, "Opportunities for Combined Heat and Power at Wastewater Treatment Facilities: Market Analysis and Lessons from the Field," (2011), https://www.epa.gov/sites/production/files/2015-07/documents/opportunities_for_combined_heat_and_power_at_wastewater_treatment_facilities_market_analysis_and_lessons_from_the_field.pdf.

9. City of Charlotte, "Combined Heat and Power Facility at McAlpine Creek Wastewater Treatment Plant," https://charlottenc.gov/newsroom/releases/Pages/McAlpineCreekWastewaterTreatmentPlant.aspx.

10. National Biosolids Partnership, "The Potential Power of Renewable Energy Generation from Wastewater and Biosolids Fact Sheet," (2014), https://www.resourcerecoverydata.org/Potential_Power_of_Renewable_Energy_Generation_From_Wastewater_and_Biosolids_Fact_Sheet.pdf.

11. R. C. Brears to Mark and Focus 2018, https://medium.com/mark-and-focus/hong-kongs-futuristic-waste-to-energy-facility-76cc0e7d4c33.

12. National Biosolids Partnership, "The Potential Power of Renewable Energy Generation from Wastewater and Biosolids Fact Sheet".

13. IKEA, "Ikea Berlin-Lichtenberg. This Is How We Save Energy" (Mainz, Germany: IKEA Verwaltungs-GmbH, 2012).

14. US EPA, "On-Site Renewable Energy Generation: A Guide to Developing and Implementing Greenhouse Gas Reduction Programs," (2014), https://www.energy.gov/sites/prod/files/2018/11/f57/onsiterenewables508.pdf.

15. Wannon Water, "Hamilton Gets Australia-First Solar Array," http://www.wannonwater.com.au/2017/november/hamilton-gets-australia-first-solar-array.aspx.

16. Marco Rosa-Clot, Giuseppe Marco Tina, and Sandro Nizetic, "Floating Photovoltaic Plants and Wastewater Basins: An Australian Project," *Energy Procedia* 134 (2017).

17. United Utilities, "Floating Solar Farm Gets Under Way at Lancaster Reservoir," https://www.unitedutilities.com/corporate/newsroom/latest-news/floating-solar-farm-gets-under-way-at-lancaster-reservoir/.

18. US EPA, "On-Site Renewable Energy Generation: A Guide to Developing and Implementing Greenhouse Gas Reduction Programs".

19. Berliner Wasserbetriebe, "Wind Turbines," http://www.bwb.de/de/7550.php.

20. National Renewable Energy Laboratory, "Energy Recovery Hydropower: Prospects for Off-Setting Electricity Costs for Agricultural, Municipal, and Industrial Water Providers and Users. July 2017–September 2017," (2018), https://www.nrel.gov/docs/fy18osti/70483.pdf.

21. R. C. Brears, January 11th, 2018, https://medium.com/mark-and-focus/the-water-energy-nexus-in-athens-4c496ac778bb.

22. Sukalyan Sengupta, Tabish Nawaz, and Jeffrey Beaudry, "Nitrogen and Phosphorus Recovery from Wastewater," *Current Pollution Reports* 1, no. 3 (2015).

23. IWA, "State of the Art Compendium Report on Resource Recovery from Water," (2016), http://www.iwa-network.org/publications/state-of-the-art-compendium-report-on-resource-recovery-from-water/.

24. Jianyin Huang et al., "Removing Ammonium from Water and Wastewater Using Cost-Effective Adsorbents: A Review," *Journal of Environmental Sciences* 63 (2018).
25. BusinessEurope, "Yara's Recovery and Reuse of Nitrogen from Municipal Waste Water," http://www.circulary.eu/project/yara-recovery/.
26. IWA, "State of the Art Compendium Report on Resource Recovery from Water".
27. City of Saskatoon, "Nutrient Recovery Facility," https://www.saskatoon.ca/services-residents/power-water/water-wastewater/wastewater/wastewater-treatment-plant/nutrient-recovery-facility.
28. Water Environment Federation, "Land Application and Composting of Biosolids," (2010), https://www.wef.org/globalassets/assets-wef/3---resources/topics/a-n/biosolids/technical-resources/wef-land-app-fact-sheet---rev0510.pdf.
29. Gryaab, "Sludge," http://www.gryaab.se/vad-vi-gor/slam/.
30. IEA Bioenergy Task 37, "Revaq Certified Wastewater Treatment Plants in Sweden for Improved Quality of Recycled Digestate Nutrients," (2015), https://www.ieabioenergy.com/wp-content/uploads/2018/01/REVAQ_CAse_study_A4_1.pdf.
31. Polymer Solutions, "Wastewater Put to Use Making Bioplastics," https://www.polymersolutions.com/blog/wastewater-put-to-use-making-bioplastics/.
32. *bioplastics MAGAZINE*, "World First—Pha from Sewage Sludge," https://www.bioplasticsmagazine.com/en/news/meldungen/20151023-Sewage-based-PHA-produced.php.
33. Deng-Fong Lin and Chih-Huang Weng, "Use of Sewage Sludge Ash as Brick Material," *Journal of Environmental Engineering* 127, no. 10 (2001).
34. Bernd Wiebusch and Carl Franz Seyfried, "Utilization of Sewage Sludge Ashes in the Brick and Tile Industry," *Water Science and Technology* 36, no. 11 (1997).
35. Federal Highway Administration U.S. Department of Transportation, "User Guidelines for Waste and Byproduct Materials in Pavement Construction," https://www.fhwa.dot.gov/publications/research/infrastructure/structures/97148/ssl.cfm.
36. New Civil Engineer, "Millions of Bricks to Be Made of Recycled Sewage Waste," https://www.newcivilengineer.com/latest/millions-of-bricks-to-be-made-of-recycled-sewage-waste/10041573.article.
37. U.S. Army Corps of Engineers, "Energy and Resource Recovery from Wastewater Treatment: State of the Art and Potential Application for the Army and the Dod," (2015), https://apps.dtic.mil/dtic/tr/fulltext/u2/a619808.pdf.
38. ZERO BRINE, "Water Plant I Netherlands," https://zerobrine.eu/pilot-projects/netherlands/.

REFERENCES

American Biogas Council. "What Is Anaerobic Digestion?" https://www.americanbiogascouncil.org/biogas_what.asp#.

Berliner Wasserbetriebe. "Wind Turbines." http://www.bwb.de/de/7550.php.

bioplastics MAGAZINE. "World First—Pha from Sewage Sludge." https://www.bioplasticsmagazine.com/en/news/meldungen/20151023-Sewage-based-PHA-produced.php.

Brears, R. C. *The Green Economy and the Water-Energy-Food Nexus.* London, UK: Palgrave Macmillan, 2017.

———. "Hong Kong's Futuristic Waste-to-Energy Facility." Mark and Focus, 2018.

———. "The Water-Energy Nexus in Athens." Mark and Focus, 2018.

BusinessEurope. "Yara's Recovery and Reuse of Nitrogen from Municipal Waste Water." http://www.circulary.eu/project/yara-recovery/.

City of Charlotte. "Combined Heat and Power Facility at McAlpine Creek Wastewater Treatment Plant." https://charlottenc.gov/newsroom/releases/Pages/McAlpineCreekWastewaterTreatmentPlant.aspx.

City of Portland. "Columbia Boulevard Wastewater Treatment Plant Renewable Natural Gas Facility Project Pub Meeting." (2017). https://www.portlandoregon.gov/cbo/article/645633.

City of Saskatoon. "Nutrient Recovery Facility." https://www.saskatoon.ca/services-residents/power-water/water-wastewater/wastewater/wastewater-treatment-plant/nutrient-recovery-facility.

Gryaab. "Sludge." http://www.gryaab.se/vad-vi-gor/slam/.

Hamburg Wasser. "The Sewage Treatment Plant Hamburg." https://www.hamburgwasser.de/privatkunden/unser-wasser/der-weg-des-wassers/abwasserreinigung/klaerwerk-hamburg/.

Huang, Jianyin, Nadeeka Rathnayake Kankanamge, Christopher Chow, David T. Welsh, Tianling Li, and Peter R. Teasdale. "Removing Ammonium from Water and Wastewater Using Cost-Effective Adsorbents: A Review." *Journal of Environmental Sciences* 63 (2018, January 1): 174–97.

IEA Bioenergy Task 37. "Revaq Certified Wastewater Treatment Plants in Sweden for Improved Quality of Recycled Digestate Nutrients." (2015). https://www.ieabioenergy.com/wp-content/uploads/2018/01/REVAQ_CAse_study_A4_1.pdf.

IKEA. "Ikea Berlin-Lichtenberg. This Is How We Save Energy." Mainz, Germany: IKEA Verwaltungs-GmbH, 2012.

IWA. "State of the Art Compendium Report on Resource Recovery from Water." (2016). http://www.iwa-network.org/publications/state-of-the-art-compendium-report-on-resource-recovery-from-water/.

Lin, Deng-Fong, and Chih-Huang Weng. "Use of Sewage Sludge Ash as Brick Material." *Journal of Environmental Engineering* 127, no. 10 (2001, October 1): 922–27.

National Biosolids Partnership. "The Potential Power of Renewable Energy Generation from Wastewater and Biosolids Fact Sheet." (2014). https://www.resourcerecoverydata.org/Potential_Power_of_Renewable_Energy_Generation_From_Wastewater_and_Biosolids_Fact_Sheet.pdf.

National Renewable Energy Laboratory. "Energy Recovery Hydropower: Prospects for Off-Setting Electricity Costs for Agricultural, Municipal, and Industrial Water Providers and Users. July 2017–September 2017." (2018). https://www.nrel.gov/docs/fy18osti/70483.pdf.

New Civil Engineer. "Millions of Bricks to Be Made of Recycled Sewage Waste." https://www.newcivilengineer.com/latest/millions-of-bricks-to-be-made-of-recycled-sewage-waste/10041573.article.

NNFCC Biocentre. "Biogas." http://www.biogas-info.co.uk/about/biogas/.

Polymer Solutions. "Wastewater Put to Use Making Bioplastics." https://www.polymersolutions.com/blog/wastewater-put-to-use-making-bioplastics/.

Rosa-Clot, Marco, Giuseppe Marco Tina, and Sandro Nizetic. "Floating Photovoltaic Plants and Wastewater Basins: An Australian Project." *Energy Procedia* 134 (2017, October 1): 664–74.

Sengupta, Sukalyan, Tabish Nawaz, and Jeffrey Beaudry. "Nitrogen and Phosphorus Recovery from Wastewater." *Current Pollution Reports* 1, no. 3 (2015, September 1): 155–66.

U.S. Army Corps of Engineers. "Energy and Resource Recovery from Wastewater Treatment: State of the Art and Potential Application for the Army and the Dod." (2015). https://apps.dtic.mil/dtic/tr/fulltext/u2/a619808.pdf.

U.S. Department of Transportation, Federal Highway Administration. "User Guidelines for Waste and Byproduct Materials in Pavement Construction." https://www.fhwa.dot.gov/publications/research/infrastructure/structures/97148/ss1.cfm.

United Utilities. "Floating Solar Farm Gets Under Way at Lancaster Reservoir." https://www.unitedutilities.com/corporate/newsroom/latest-news/floating-solar-farm-gets-under-way-at-lancaster-reservoir/.

US EPA. "On-Site Renewable Energy Generation: A Guide to Developing and Implementing Greenhouse Gas Reduction Programs." (2014). https://www.energy.gov/sites/prod/files/2018/11/f57/onsiterenewables508.pdf.

———. "Opportunities for Combined Heat and Power at Wastewater Treatment Facilities: Market Analysis and Lessons from the Field." (2011). https://www.epa.gov/sites/production/files/2015-07/documents/opportunities_for_combined_heat_and_power_at_wastewater_treatment_facilities_market_analysis_and_lessons_from_the_field.pdf.

Wannon Water. "Hamilton Gets Australia-First Solar Array." http://www.wannonwater.com.au/2017/november/hamilton-gets-australia-first-solar-array.aspx.

Waste Management World. "Food Waste to Be Co-digested into Biogas in Korean Underground Wastewater Treatment Plant." https://waste-management-world.com/a/food-waste-to-be-co-digested-into-biogas-in-korean-underground-wastewater-treatment-plant.

Water Environment Federation. "Land Application and Composting of Biosolids." (2010). https://www.wef.org/globalassets/assets-wef/3—resources/topics/a-n/biosolids/technical-resources/wef-land-app-fact-sheet—rev0510.pdf.

———. "Water Environment Federation Position Statement: Renewable Energy Generation from Wastewater." (2011). https://www.wef.org/globalassets/assets-wef/5—advocacy/policy-statements/policy-statements/wef-position-statementonrenewable-energybotfinal14oct2011-1.pdf.

Wiebusch, Bernd, and Carl Franz Seyfried. "Utilization of Sewage Sludge Ashes in the Brick and Tile Industry." *Water Science and Technology* 36, no. 11 (1997, January 1): 251–58.

ZERO BRINE. "Water Plant I Netherlands." https://zerobrine.eu/pilot-projects/netherlands/.

CHAPTER 8

Anglian Water Developing the Circular Water Economy

Abstract Anglian Water has implemented a range of technology policies to facilitate the development of the circular water economy that not only mitigates greenhouse gas emissions but also enhances resilience to climate change.

Keywords Water conservation · Water efficiency · Water reuse · Water recycling · Resource recovery · Renewable energy

INTRODUCTION

Anglian Water supplies water and water recycling services to more than six million domestic customers in the east of England and Hartlepool. Over the past 20 years, the region's population has grown by 20%, but the utility still provides the same amount of water today as they had done so in 1990 by minimising leaks and encouraging more waterwise customers. Anglian Water is the largest water and water recycling company in England and Wales by geographic area and services one of the driest regions in the country, with just 600 millimetres of rain each year, on average a third less than the rest of England.[1] The utility is also one of the largest energy users in the East of England, resulting in significant greenhouse gas emissions. This chapter will discuss how Anglian Water has implemented a range of 3R technology policies (reduce, reuse

R. C. Brears, *Developing the Circular Water Economy*,
Palgrave Studies in Climate Resilient Societies,
https://doi.org/10.1007/978-3-030-32575-6_8

and recycle, and recover) that facilitate the development of the circular water economy that not only mitigates greenhouse gas emissions but also enhances resilience to climate change.

DEVELOPING THE CIRCULAR WATER ECONOMY: REDUCE

Anglian Water has implemented a range of technology policies to enhance water conservation and promote water efficiency measures in the development of the circular water economy.

Water-Saving Home Visits and Kits

Anglian Water provides water-saving home visits in which a member of the utility will provide water-saving advice and fit water-saving products where possible. The utility also offers a free garden water-saving kit to customers to conserve water while still having a beautiful garden. The water-saving products and kits freely distributed are summarised in Table 8.1.[2,3]

Purpose-Built Education Centres

Anglian Water has purpose-built education centres at its water recycling centres at Chelmsford in Essex and Leighton Linslade in Bedfordshire for school visits. Visits involve a tour where students can discover the sights, sounds, and smells of a water recycling centre. Fun, interactive sessions in a classroom provide a range of experiences that bring water topics to life. Sessions are led by a qualified teacher and are designed to link to the school curriculum.[4]

Smart Meter Trials

Anglian Water is implementing smart meter trials in Colchester, Newmarket, and Norwich.

Colchester

Smart meters were installed in parts of Colchester in 2015 with the signal read by a reader fitted to Colchester Borough Council's refuse collection vehicles. This partnership between Anglian Water and Colchester

Table 8.1   Anglian Water's free water-saving products

Water-saving programme	Product	Description
Water-saving home visit	Dual flush converter	Reduces the water used with every flush by up to 50%
	Tap inserts	Saves customers money every time the tap is turned on
	Eco Pulse shower head	Achieves the temperature and powerful flow of traditional shower heads with only 40–60% of the water by creating gaps between droplets
	Showersave	An adapter that fits to the shower changing it to a constant 7.6 litres per minute flow
	Digital shower timer	Every minute saved in the shower could save up to 10,000 litres of water per year
	Hosepipe gun	Save hundreds of litres of water a year
Free garden water saving kit	Water storing crystals	The crystals are mixed in compost and when water is added the granules are activated and expand into thousands of self-contained 'reservoirs', each absorbing hundreds of times their own weight of water. These can reduce watering frequencies by 75%
	Water storing mats	These can be used in hanging baskets, tubs, and planters. They give a permanently stable water reservoir within compost
	Seed sticks	A selection of basil, thyme, chives, and parsley seeds for your own herb garden
	Water in the garden leaflet	A leaflet on saving water, time, and money to keep a garden green and growing
	Pocket guide to drought-tolerant plants	Helps choose the right plants to cope with dry conditions

Borough Council means that most properties' meters are read weekly, with meter readings available to customers on their 'My Use' portal.

Newmarket

A new generation of smart meters is being installed in Newmarket with Anglian Water working to install 7500 smart meters throughout the town. The smart meters transmit hourly readings from the property with customers able to access it via their 'My Use' portal. This helps customers to see their water use in more detail and over time will allow them to make comparisons between their water use from year to year and between similar types of properties in the area. In addition to accessing their water usage, customers will also be able to access free tips on how to save water, make pledges to change their water behaviour, and track the effect of the change on their water use.

Norwich
The next generation of smart meters is being installed in North Norwich and surrounding areas. Anglian Water plans to install 11,000 smart meters that transmit hourly readings from the property. The meter readings are shared with customers on a daily basis through the online service 'My Use'.[5]

Anglian Water Business Services for Industry

Anglian Water Business is a national water retailer, supplying more than 140,000 organisations across the United Kingdom. The organisation provides a variety of services to customers helping them make sustainable savings that benefit both the business and the environment, including:

Water Management Strategy
Anglian Water Business can work with customers to develop an effective water management strategy to identify and address any water risks that could impact operational efficiency. Specifically, the utility can help businesses:

- Develop an understanding of their environmental impact
- Identify and manage their water risk
- Manage water risk across the outsourced supplier base
- Create an effective water management plan.

Automatic Meter Reading Service
Anglian Water Business offers customers an Automatic Meter Reading (AMR) Service that utilises the latest AMR technologies. The service is offered on a five-year term and includes:

- The supply, installation, and maintenance of the latest AMR device
- A five-year warranty for the hardware
- Flexible access for customers to WaterCore, Anglian Water Business' online data presentment and analytics system that provides data

downloads in Excel format. Data can be presented in 15-minute intervals

- The ability to set automated/instant alerts against customer-defined usage.

Active Water Management®

Anglian Water Business can analyse for customers their water usage through its Active Water Management® programme which alerts clients of opportunities for reducing the amount of water they use. Anglian Water Business reviews water usage data for its clients and detects any irregular consumption that may indicate inefficient practices, faulty equipment or leaks.

Water Efficiency Audits

Anglian Water Business offers water efficiency audits that assess domestic water use on site, such as toilets, taps, kitchens, and cleaning facilities, and identify ways to save water. The audit involves a visit by one of the utility's water efficiency assessors who will review the organisation's current water consumption and specific aspects of the customer's water installation including:

- Analysing of meter readings to determine current water consumption patterns
- Identifying site sub-metering
- Assessing possible leakage above and below ground
- A general inspection of domestic water systems
- An assessment of urinal cistern control mechanisms
- Feedback on any water management procedures the customer may need to consider
- Summary of the potential water, carbon, and water-related energy savings that could be made.

Process Water Efficiency Audits

Anglian Water Business can help customers identify potential optimisation opportunities and water savings across all site-based production and operational processes. This will involve one of the utility's industrial engineers conducting a thorough review, engaging with the site's employees to ensure the conclusions are informed and recommendations are commercially viable. The utility will then work with the business customer to determine the next steps.[6]

Anglian Water Business Services for Agriculture

To ensure the agricultural sector can reduce costs, improve operational efficiency, maintain compliance, and secure future resilience, Anglian Water Business offers a range of services including:

- *Water auditing and benchmarking*: Anglian Water Business can conduct a water audit to establish where, when, and how much water is used and what can be done to improve water efficiency. The audit can also be used as a benchmark to assess performance against that of other agricultural businesses
- *Water strategy*: Anglian Water Business can help customers understand the risks of water scarcity, how to plan and prepare for scarcity while maintaining profitability, and how to invest over extended timescales to protect the business.[7]

Reducing Leakage

The utility's Optimised Water Networks Strategy aims to prevent bursts and leaks through better management of the pressure in the network. In 2012, the Peterborough Network Calming programme installed three large-diameter pressure-reducing valves and three major cross connections to manage water pressure in the central area of Peterborough. This allows water flowing through the mains to be automatically controlled and adjusted in response to changing demand from customers. Since then the scheme has been extended to enable the proactive management of pressure to 90% of connected properties in the Peterborough area, enabling Anglian Water to see:

- A total drop of 5.87 million litres per day (Ml/day) in water lost to leaks and bursts
- An overall fall of 6 Ml/day in the amount of water supplied
- An overall reduction of more than 50 burst mains and more than 200 other leaks every year.[8]

DEVELOPING THE CIRCULAR WATER ECONOMY: REUSE AND RECYCLE

Anglian Water has implemented a range of technology policies to increase the reuse and recycling of water resources in the development of the circular water economy.

Green Water

Exploring ways to reduce water use, as well as new and sustainable ways to supply water, is a priority for Anglian Water. As such, the utility is encouraging developers to install Green Water systems in new homes, which is fit for purpose water for household appliances where drinking water quality is not necessary, such as flushing toilets, watering gardens, and for use in washing machines. Anglian Water is prioritising the development of:

- *Rainwater harvesting systems:* Capturing of rainfall directly from the roof by redirecting the downpipes into a storage tank
- *Stormwater harvesting systems:* Capturing surface water runoff in a storage tank or pond. The water can be treated if required then supplied to the house through a green pipe network
- *Water recycling systems:* Capturing and treating the used water from homes for reuse. This can include options for individual houses, for example, greywater reuse, or much larger centralised water recycling schemes.[9]

Water Efficiency Incentive

The Water Efficiency Incentive aims to help reduce water use in new homes by encouraging homebuilders to build sustainable homes that incorporate water efficiency measures, including Green Water systems. Developers building properties to a water efficiency standard of 100 litres per person per day receive a waiver of the fixed element of the Zonal Charge, resulting in potential savings of up to £740 per plot. The process is summarised in Fig. 8.1.[10]

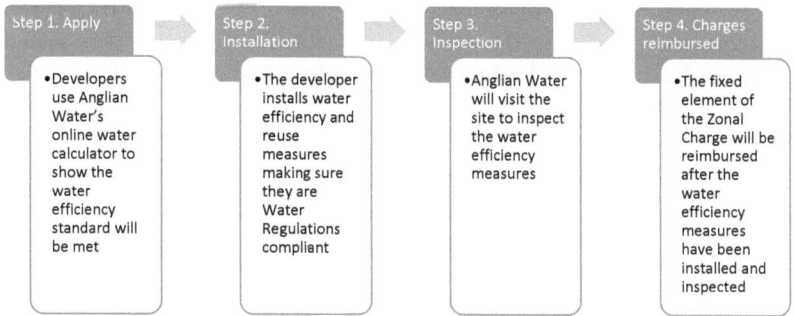

Fig. 8.1 Anglian Water's water efficiency incentive

Natural Treatment Plant

Anglian Water, in partnership with Norfolk Rivers Trust, has created a natural treatment plant for over a million litres of water a day to help improve the quality of water that is returned to the River Ingol, which is a spring-fed chalk stream. Used but treated water passes through the wetland to be further filtered and cleaned by the wetland plants before it is returned to the river. In addition, the wetland is a biodiversity asset for the region as it attracts breeding birds, amphibians, bats, and water voles to the local environment. The £500,000 project is made up of four shallow interconnected ponds which have been planted with native chalk wetland species such as iris, sedges, rush, marsh marigold, and watercress. The plants naturally clean the water, removing ammonia and phosphate before it goes back into the river. Anglian Water's existing treatment plants already remove the majority of these substances in line with environmental permits issued by the Environment Agency, but the wetland filters it further to ensure it is of an even higher standard, removing the need for expensive, high carbon treatment equipment conventionally used.[11,12]

DEVELOPING THE CIRCULAR WATER ECONOMY: RECOVER

Anglian Water has implemented a range of technology policies to recover resources in the development of the circular water economy.

Combined Heat and Power

Anglian Water has 10 existing sludge treatment centres and Combined Heat and Power (CHP) engines. At the utility's Cambridge Water Recycling Centre much of the site's power is met using renewable energy with the CHP engines providing efficiency savings in both cost and carbon footprint. Anglian Water is trialling the use of E-STOR technology that uses second life Renault electric vehicle batteries to provide a smart, affordable, and flexible approach to grid load management. Operating software allows the centre to maximise efficiencies and savings by storing excess energy for when it is needed such as times of peak demand or when energy tariffs are high.[13]

Solar Energy

Anglian Water has identified a number of opportunities to generate renewable energy from solar panels including the following:

Solar Power and Storage

Anglian Water is working with commercial partners on an integrated energy storage project that will see a 60 kW/300 kWh energy storage machine installed alongside a 450 kWp solar photovoltaic (PV) system at one of its water treatment works. The energy storage machine will provide Anglian Water with real-time balancing services to take advantage of wholesale energy price arbitrage. In total, the project is expected to reduce site electricity costs by 50% per annum by 2040. Overall, the project will enable the utility to store excess solar generated during the day and use it at other times.

Over the next 18 months, Anglian Water will be building over 30 MWp of solar under a 25-year Power Purchase Agreement with corporate partners. The programme will reduce Anglian Water's carbon emissions by 15,000 tonnes of carbon dioxide equivalent and increase renewable energy generation by around 25%, delivering annual savings of over £1 million.[14]

Solar Energy at Grafham Water

Grafham Water is a key operational site for water treatment and distribution, using around 45 million kilowatt-hours of energy a year to supply clean water. To reduce energy usage and lower carbon emissions, Anglian Water will install solar panels on land at the site to generate over a quarter of the energy used by the site per annum, which in turn will save 4500 tonnes of carbon each year.[15]

Wind Turbines

Anglian Water currently has three wind turbines that generate around 14 GWhs of energy per annum. The utility is continuing to explore wind power in places where it would be appropriate.[16]

Case Study Summary

Anglian Water has implemented a range of 3R technology policies that facilitate the development of the circular water economy that not only mitigates greenhouse gas emissions but also enhances resilience to climate change (Table 8.2).

Table 8.2   Anglian Water case study summary

3R principle	Policy innovation	Description	
Reduce	Water-Saving Home Visit and Kits	Anglian Water provides water-saving home visits in which a member of the utility will provide water-saving advice and fit water-saving products where possible	
	Free garden water saving kit	The utility offers a free garden water saving kit to customers to conserve water while still having a beautiful garden	
	Purpose-Built Education Centres	Anglian Water's purpose-built education centres at its water recycling centres provide an informative educational experience for students on water conservation	
	Smart Meter Trials	Anglian Water is trialling smart meters in several communities, with the smart meters transmitting hourly readings from the property with customers able to access it via their 'My Use' portal	
	Anglian Water Business Services for Industry	Water Management Strategy	Anglian Water Business can work with customers to develop an effective water management strategy to identify and address any water risks that could impact operational efficiency
		Automatic Meter Reading Service	Anglian Water Business offers customers an AMR Service
		Active Water Management	Anglian Water Business can analyse for customers their water usage
		Water Efficiency Audits	Utility representative assesses domestic water use on site and identifies ways to save water
		Process Water Efficiency Audits	Helps customers identify potential optimisation opportunities and water savings across all site-based production and operational processes
	Anglian Water Business Services for Agriculture	Water auditing and benchmarking	Anglian Water Business can conduct a water audit to improve water efficiency. The audit can be used as a benchmark to assess performance against other agricultural businesses
		Water strategy	Anglian Water Business can help customers understand the risks of water scarcity and protect their business
	Reducing Leakage	The Optimised Water Networks Strategy aims to prevent bursts and leaks through better management of the pressure in the network	
Reuse and recycle	Green Water	The utility is encouraging developers to install Green Water systems in new homes including rainwater harvesting systems, stormwater harvesting systems, and water recycling systems	
	Water Efficiency Incentive	Homebuilders receive a waiver on zonal charges for building sustainable homes that incorporate water efficiency measures, including Green Water systems	
	Natural Treatment Plant	Anglian Water, in partnership with Norfolk Rivers Trust, has created a natural treatment plant for over a million litres of water a day to help improve the quality of water that is returned to the River Ingol, which is a spring-fed chalk stream. Used but treated water passes through the wetland to be further filtered and cleaned by the wetland plants before it is returned to the river. In addition, the wetland is a biodiversity asset for the region as it attracts breeding birds, amphibians, bats, and water voles to the local environment	

(continued)

Table 8.2 (continued)

3R principle	Policy innovation	Description	
Recover	Combined Heat and Power	Anglian Water operates CHP engines. It is also trialling the use of E-STOR technology that uses second life Renault electric vehicle batteries to provide a smart, affordable, and flexible approach to grid load management	
	Solar Energy	Solar Power and Storage	Anglian Water is working with commercial partners on an integrated energy storage project that will see an energy storage machine installed alongside a solar PV system at a water treatment works
		Solar Energy at Grafham Water	Anglian Water will install solar panels on land at the site to generate over a quarter of the energy used by the site per annum
	Wind Turbines	Anglian Water currently has three wind turbines installed and will continue to seek appropriate sites for future wind turbines	

NOTES

1. Anglian Water, "Our Company," https://www.anglianwater.co.uk/about-us/our-company.aspx.
2. "The Big Bits & Bobs Giveaway," https://www.anglianwater.co.uk/environment/how-you-can-help/using-water-wisely/we-products/products.aspx.
3. "Welcome to the Potting Shed Club," https://www.anglianwater.co.uk/environment/how-you-can-help/using-water-wisely/save-water-in-the-garden.aspx.
4. "Education Centres," https://www.anglianwater.co.uk/community/education/education-centres/.
5. "Water Smart Meters," https://www.anglianwater.co.uk/household/water-meters/smart-meters/.
6. Anglian Water Business, "Water Efficiency," https://www.anglianwater-business.co.uk/services/water/water-efficiency/.
7. "Agriculture," https://www.anglianwaterbusiness.co.uk/your-sector/agriculture/.
8. Anglian Water, "Prevention Is Better Than Cure," https://www.anglian-water.co.uk/about-us/annual-reports/own.aspx.
9. "Green Water," https://www.anglianwater.co.uk/developers/green-water.aspx.
10. "Water Efficiency Incentive," https://www.anglianwater.co.uk/developers/water-efficiency-incentive.aspx.

11. "Norfolk Wetland Hailed a Success as Anglian Water Outlines Plans for £800Million of Environmental Investment," https://www.anglianwater.co.uk/news/norfolk-wetland-hailed-a-success-as-anglian-water-outlines-plans-for-800million-of-environmental-investment/.
12. Catchment Based Approach, "Norfolk Rivers Trust Create Wetland Water Treatment Facility for Anglian Water," https://catchmentbasedapproach.org/learn/norfolk-rivers-trust-create-wetland-water-treatment-facility-for-anglian-water/.
13. Anglian Water, "Driving Efficiency for the Future," https://media.anglianwater.co.uk/driving-efficiency-for-the-future/.
14. WaterBriefing, "Anglian Water Solar Power & Storage Scheme Will up Energy Generation by 80% & Create Extra Revenue Stream," https://www.waterbriefing.org/home/technology-focus/item/15383-anglian-water-partnership-will-integrate-solar-and-storage-at-uk-water-treatment-facilities.
15. Anglian Water, "Anglian Water Takes Another Step Towards Its Goal to Be a Carbon Neutral Business by 2050," https://www.anglianwater.co.uk/news/anglian-water-takes-another-step-towards-its-goal-to-be-a-carbon-neutral-business-by-2050/.
16. "Renewable Energy," https://www.anglianwater.co.uk/environment/why-we-care/renewable-energy.aspx.

References

Anglian Water. "Anglian Water Takes Another Step Towards Its Goal to Be a Carbon Neutral Business by 2050." https://www.anglianwater.co.uk/news/anglian-water-takes-another-step-towards-its-goal-to-be-a-carbon-neutral-business-by-2050/.

———. "The Big Bits & Bobs Giveaway." https://www.anglianwater.co.uk/environment/how-you-can-help/using-water-wisely/we-products/products.aspx.

———. "Driving Efficiency for the Future." https://media.anglianwater.co.uk/driving-efficiency-for-the-future/.

———. "Education Centres." https://www.anglianwater.co.uk/community/education/education-centres/.

———. "Green Water." https://www.anglianwater.co.uk/developers/green-water.aspx.

———. "Norfolk Wetland Hailed a Success as Anglian Water Outlines Plans for £800Million of Environmental Investment." https://www.anglianwater.co.uk/news/norfolk-wetland-hailed-a-success-as-anglian-water-outlines-plans-for-800million-of-environmental-investment/.

———. "Our Company." https://www.anglianwater.co.uk/about-us/our-company.aspx.

———. "Prevention Is Better Than Cure." https://www.anglianwater.co.uk/about-us/annual-reports/own.aspx.

———. "Renewable Energy." https://www.anglianwater.co.uk/environment/why-we-care/renewable-energy.aspx.

———. "Water Efficiency Incentive." https://www.anglianwater.co.uk/developers/water-efficiency-incentive.aspx.

———. "Water Smart Meters." https://www.anglianwater.co.uk/household/water-meters/smart-meters/.

———. "Welcome to the Potting Shed Club." https://www.anglianwater.co.uk/environment/how-you-can-help/using-water-wisely/save-water-in-the-garden.aspx.

Anglian Water Business. "Agriculture." https://www.anglianwaterbusiness.co.uk/your-sector/agriculture/.

———. "Water Efficiency." https://www.anglianwaterbusiness.co.uk/services/water/water-efficiency/.

Catchment Based Approach. "Norfolk Rivers Trust Create Wetland Water Treatment Facility for Anglian Water." https://catchmentbasedapproach.org/learn/norfolk-rivers-trust-create-wetland-water-treatment-facility-for-anglian-water/.

WaterBriefing. "Anglian Water Solar Power & Storage Scheme Will up Energy Generation by 80% & Create Extra Revenue Stream." https://www.water-briefing.org/home/technology-focus/item/15383-anglian-water-partnership-will-integrate-solar-and-storage-at-uk-water-treatment-facilities.

CHAPTER 9

Austin Water Developing the Circular Water Economy

Abstract Austin Water has implemented a range of technology policies to facilitate the development of the circular water economy that not only mitigates greenhouse gas emissions but also enhances resilience to climate change.

Keywords Water conservation · Water efficiency · Water reuse · Water recycling · Resource recovery · Renewable energy

INTRODUCTION

The City of Austin receives its municipal water supply from the Colorado River. With Austin being one of the fastest-growing cities in the United States, Austin Water will be challenged by increased demand on already scarce resources. Regarding climate change, the Austin region will see longer and deeper periods of drought, along with periods of heavy rain events. To mitigate emissions, the city's Climate Program aims to achieve carbon neutrality for City of Austin operations by 2020.[1,2] This chapter will discuss how Austin Water has implemented a range of 3R technology policies (reduce, reuse and recycle, and recover) that facilitate the development of the circular water economy that not only mitigates greenhouse gas emissions but also enhances resilience to climate change.

R. C. Brears, *Developing the Circular Water Economy,*
Palgrave Studies in Climate Resilient Societies,
https://doi.org/10.1007/978-3-030-32575-6_9

Developing the Circular Water Economy: Reduce

Austin Water has implemented a range of technology policies to enhance water conservation and promote water efficiency measures in the development of the circular water economy.

Water Rates

Austin Water has a five-tier fixed charge for residential customers, based on the total billed water consumption for the billing period, and a five-tier volume charge, which is the rate charged per 1000 gallons of total billed water consumption for the billing period (Table 9.1).[3]

Free Water Conservation Tools: Residents

Austin Water offers a variety of free tools to help residents conserve water including:

Dropcountr Inc. Mobile and Web Application

Austin Water has contracted Dropcountr Inc. to offer free, digital home water use reports. The reports help customers save both water and money. Reports are available by mobile app and/or by the Internet and include:

- A customised household water use profile
- Information about a customer's past water use compared to:
  - Similar households
  - Utility bill rate tiers

Table 9.1 Austin Water's water rates

Gallons of water	Fixed charge	Volume charge
0–2000	$1.25	$2.89
2001–6000	$3.55	$4.81
6001–11,000	$9.25	$8.34
11,001–20,000	$29.75	$12.70
20,001+	$29.75	$14.21

- Water efficiency standards
- The customer's water savings goal
- Suggestions for ways to save water and links to Austin Water conservation goals
- Utility alerts and announcements about new conservation programmes.[4]

Indoor and Outdoor Water Conservation Tools

Austin Water offers free tools to help residential customers save water both indoors and outdoors, summarised in Table 9.2.[5]

Irrigation System Evaluation

Irrigation system evaluations can help residents set an efficient watering schedule and identify the need for system repairs and upgrades. Customers can schedule a Free Irrigation System Evaluation by a licenced irrigator from Austin Water if they have an in-ground sprinkler system and have used more than 25,000 gallons in one month or more than 20,000 gallons in two consecutive months during the irrigation season (April–October).[6]

Table 9.2 Free residential indoor and outdoor water conservation tools

Indoor/outdoor	*Tool*	*Description*
Indoor	Efficient showerhead	Saves two gallons per minute or more with a water-efficient showerhead in either regular or soap-up valve models that only use 1.5 gallons of water per minute
	Kitchen and bathroom faucet aerator	Bathroom aerators use 0.5 gallons of water per minute and kitchen aerators use 1.5 gallons per minute
Outdoor	Soil moisture meter	Measure the moisture of soil to help know when it's time to add water
	Water saver hose meter	Monitor and control water use with a digital meter attachment for garden hoses and hose-end sprinklers
	Sunlight calculator	Measure the amount of light each area of the yard receives then use the included plant guide to select plants that best fit the light conditions

Free Water Conservation Tools: Commercial/Multi-family/Schools

Austin Water offers a variety of free tools to help commercial/multi-family/school customers conserve water, examples of which include:

3C Business Challenge

Austin Water's 3C Business Challenge encourages businesses to learn ways of reducing water use by 10%, which in turn can lower energy and wastewater costs as well. Participating businesses complete an application and self-audit form which is then assessed by an Austin Water representative who will make recommendations on how to improve water efficiency and suggestions on rebates to help with costs for water-efficient upgrades.[7]

Free Showerhead and Faucet Aerators: Multi-family Customers Only

Austin Water's multi-family customers can participate in Austin Energy's multi-family water and energy packaged rebates programme which includes the free distribution of water-efficient showerheads and faucet aerators.[8]

WaterWise Hotel Partnership

Austin Water's WaterWise Hotel Partnership offers free recognition for lodging facilities that use water-efficient measures and practices. Water savings can lower operating costs and enhance performance and serve as an effective marketing tool to attract guests and keep established ones. Programme participants must complete the 3C Business Challenge self-audit and submit it to Austin Water and implement all the required measures. Programme participants will then receive signage from Austin Water recognising the facility as a WaterWise Partner for display in a public area of the property and be listed as a WaterWise Partner on the Austin Water Conservation website.[9]

Conservation Restrictions

When conservation restrictions are implemented during drought conditions, watering day(s) are implemented for residential and commercial/multi-family/public schools (Table 9.3). In addition, a range of restrictions for general water use apply, with violations of water restrictions potentially leading to a fine of $500 per violation:

Table 9.3 Conservation stage restrictions

Customer type	Water usage	Restrictions
Residential	Hose-end sprinklers	Two days per week—midnight to 10 a.m. and/or 7 p.m. to midnight
	Automatic irrigation	One day per week—midnight to 10 a.m. and/or 7 p.m. to midnight (residential customers may also water a second day with a hose-end sprinkler)
Commercial/multi-family/public schools	Hose-end sprinklers or automatic irrigation	One day per week—midnight to 8 a.m. and/or 7 p.m. to midnight

- Wasting water is prohibited
- Washing vehicles at home is permitted with an auto shut-off hose or bucket
- Fountains must recirculate water
- Restaurants may not serve water unless requested by a customer
- Commercial power/pressure washing equipment must meet efficiency requirements.[10]

Austin Water Regulatory Programmes

Austin Water has several water conservation regulations that apply to commercial, multi-family, and City of Austin properties including:

Commercial Facility Irrigation Assessments

Commercial, multi-family, and City of Austin properties that are one-acre in size or larger must complete an irrigation system inspection every two years. An Austin Water Authorized Irrigation Inspector must perform the inspection.[11]

Commercial Vehicle Wash Facility Efficiency Assessments

Commercial, multi-family, and city municipal facilities with vehicle wash equipment that use potable water supplied by Austin Water must submit an annual efficiency evaluation. A plumber licenced by the State of Texas must perform the evaluation.[12]

Table 9.4 Austin Water rebates

Customer type	Rebate	Description
Residents	Irrigation upgrade	Up to $400 to improve irrigation efficiency
	Landscape survival tool	Up to $120 for compost, mulch, and core aeration service
	Pool cover	Up to $200 for new pool cover
	Pressure regulating valve	Up to $100 for pressure regulating valve
	Rainwater harvesting	Up to $5000 for equipment to capture rainwater
	Watering timer	Up to $40 for hose timers
	WaterWise landscape	Up to $1750 to convert turf grass to native beds
	WaterWise rainscape	Up to $500 for landscape features to retain rainwater
Commercial/ multi-family/school	Bucks for business	Up to $100,000 for equipment and process efficiency upgrades
	Commercial kitchen	From $40 to $2500 per item for water-efficient kitchen equipment
	Irrigation system improvement	The Irrigation System Improvement Rebate is available to increase irrigation efficiency
	Rainwater harvesting	Up to $5000 for equipment to capture rainwater
	Water efficiency audit	Up to $5000 for an audit to identify potential water savings
	Pressure regulating valve *(multi-family only)*	Up to $500 for pressure regulating valve
	WaterWise landscape *(multi-family, Home Owner Association only)*	Up to $5000 to convert turf grass to native plants
	WaterWise rainscape *(schools only)*	Up to $500 for landscape features to retain rainwater

Cooling Tower Efficiency Programme

All properties with cooling towers must register them with Austin Water so potential water-saving upgrades can be identified and owners kept informed of available rebates. All properties must also submit annual inspection forms that are performed by an independent third-party Texas licenced mechanical or chemical engineer or another licenced inspector.[13]

Rebates

Austin Water offers residents and commercial/multi-family/school customers a variety of rebates, summarised in Table 9.4.[14]

DEVELOPING THE CIRCULAR WATER ECONOMY: REUSE AND RECYCLE

Austin Water has implemented a range of technology policies to increase the reuse and recycling of water resources in the development of the circular water economy.

Alternative Water Sources

Austin Water's alternative water sources include highly treated reclaimed water from its wastewater treatment plants and onsite water sources.

Reclaimed Water

Reclaimed water is less expensive to use or treat, and the price is usually around one-third of the price of drinking water. It can be used to meet over 90% of the criteria for drinking water and has no noticeable odour and is harmless to humans through normal contact. City Code requires all new commercial developments or redevelopments within 250 feet of a reclaimed water main to connect for irrigation, cooling, and other significant non-potable water uses.[15]

Onsite Water Reuse Systems

Austin Water provides Internet resources for customers considering condensate water, greywater, rainwater, stormwater or other non-sewage originated waters on their property (onsite) and reusing them for non-potable uses. The various alternative onsite water sources that can

Table 9.5   Onsite water reuse systems

Type of system	Description
Condensate water	Water produced in heating, ventilation, and air conditioning systems as a result of evaporative cooling
Greywater	Wastewater from showers, bathtubs, handwashing lavatories, sinks used for the disposal of household or domestic products, sinks that are not used for food preparation or disposal, and clothes washing machines
Rainwater	Precipitation collected from roof surfaces or other above-ground structures
Stormwater	Precipitation collected at or below-ground surfaces
Other non-sewage originated water	Foundation drain water, swimming pool backwash and drain water, and residential reverse osmosis reject water

be used are summarised in Table 9.5. Because permitting of an onsite water reuse system in the city involves coordination between multiple City departments, Austin Water provides help to property owners in navigating the permitting process.[16]

Developing the Circular Water Economy: Recover

Austin Water has implemented a range of technology policies to recover resources in the development of the circular water economy.

Biogas to Power Wastewater Treatment Plant

All of Austin's sewage solids are pumped to the Hornsby Bend Wastewater Treatment Plant, where water from the biosolids treatment process and stormwater from the composting and basin area is treated through a pond system, with no water discharge from the site. Instead, the water is recycled for irrigation of onsite hay fields. Methane gas produced in the treatment process is recycled to generate electricity and heat. The biogas generator can generate 700 kilowatts of power, more than the 500 kilowatts needed to run the treatment plant. The excess electricity produced is fed back into the electric grid, enabling Austin

Water to receive a credit on its electric bill. The clean energy generated reduces carbon dioxide emissions by around 2,800 tonnes or the equivalent of avoiding five million vehicle miles travelled in Austin.[17,18]

Dillo Dirt

Each year, thousands of tonnes of biosolids are anaerobically digested and composted into an Environmental Protection Agency (EPA)-certified soil conditioner called Dillo Dirt™. The product is donated to landscape public spaces and sold to commercial vendors. The City of Austin has been making Dillo Dirt™ since 1989 with yard trimmings collected kerbside across the city combined with treated sewage sludge and composted to create Dillo Dirt™.

Dillo Dirt Treatment Process
Austin Water's Hornsby Bend Biosolids Management Plant (Hornsby Bend) is a nationally recognised biosolids recycling facility. Biosolids (primary and secondary waste-activated sludge) from the city's wastewater treatment plants are pumped to Hornsby Bend. They are then thickened by gravity belt thickeners to reduce the volume and allow more time for decomposition in the digesters. Once thickened, biosolids undergo anaerobic digestion with the methane gas used to power on-site electric generators. After digestion, biosolids are thickened again by belt presses before being re-used in land application and composting. A portion of the dried biosolids is combined with yard trimmings and composted. After a month, the compost is 'cured' for several months, then screened to produce Dillo Dirt™. Dillo Dirt™ meets all Texas and EPA requirements for 'unrestricted' use, including for vegetable gardens.[19]

CASE STUDY SUMMARY

Austin Water has implemented a range of 3R technology policies that facilitate the development of the circular water economy that not only mitigates greenhouse gas emissions but also enhances resilience to climate change (Table 9.6).

Table 9.6 Austin Water case study summary

3R principle	Policy innovation	Description
Reduce	Water rates	Austin Water has a five-tier fixed charge for residential customers and a five-tier volume charge
	Free water conservation tools: Residents	Dropcountr Inc. Mobile and Web Application
		Austin Water has contracted Dropcountr Inc. to offer customers free, digital home water use reports
	Indoor and Outdoor Water Conservation Tools	Austin Water offers free tools to help residential customers save water both indoors and outdoors
	Irrigation System Evaluation	Customers can have a free irrigation system evaluation to become more efficient and identify any need for repairs and upgrades
	Free water conservation tools: Commercial/ multi-family/schools	3C Business Challenge
		Encourages businesses to learn ways of reducing water use by 10%, which can lower energy and wastewater costs
	Free Showerhead and Faucet Aerators	Austin Energy's multi-family water and energy packaged rebates programme offers free water-efficient showerheads and faucet aerators
	WaterWise Hotel Partnership	Offers free recognition for lodging facilities that use water-efficient measures and practices
	Conservation restrictions	When conservation restrictions are implemented during drought conditions, watering day(s) are implemented for residential and commercial/multi-family/public schools, with violations of water restrictions potentially leading to fines

3R principle	Policy innovation	Description	
	Austin Water regulatory programmes	Commercial Facility Irrigation Assessments	Large commercial, multi-family, and City of Austin properties must complete an irrigation system inspection every two years
		Commercial Vehicle Wash Facility Efficiency Assessments	Commercial, multi-family, and city municipal facilities with vehicle washes using potable water must submit an annual efficiency evaluation
		Cooling Tower Efficiency Programme	All properties with cooling towers must register them with Austin Water so potential water-saving upgrades can be identified and owners kept informed of available rebates
	Rebates		Austin Water offers residents and commercial/multi-family/school customers a variety of rebates
Reuse and recycle	Alternative water sources	Reclaimed Water	The price of reclaimed water is around one-third of the price of drinking water. City Code requires all new commercial locations within 250 feet of a reclaimed water main to connect for non-potable water uses
		Onsite Water Reuse Systems	Austin Water provides Internet resources for customers considering onsite water reuse systems. Austin Water also provides help to property owners in the permitting process
Recover	Biogas to power		All of Austin's sewage solids are pumped to the Hornsby Bend Wastewater Treatment Plant. Methane gas produced in the treatment process is recycled to generate electricity and heat. The biogas generator can generate more power than is needed to run the treatment plant. The excess electricity produced is fed back into the electric grid
	Dillo Dirt		Biosolids are anaerobically digested and composed into an EPA-certified soil conditioner called Dillo Dirt™. The product is donated to landscape public spaces and sold to commercial vendors

NOTES

1. Austin Water, "A Water Plan for the Next 100 Years," (2018), http://austintexas.gov/sites/default/files/files/Water/WaterForward/Water_Forward_Booklet.pdf.
2. City of Austin Office of Sustainability, "Climate Change," https://www.austintexas.gov/climate.
3. Austin Water, "2017–2018 Water & Wastewater Rates," (2018), http://www.austintexas.gov/sites/default/files/files/Water/Rates/ResidentialPublicRates_05.2018.pdf.
4. "Dropcountr Home Water Use Report Frequently Asked Questions," (2019, January 9), http://www.austintexas.gov/sites/default/files/files/Water/Conservation/DropcountrFAQ.pdf.
5. "Free Water Conservation Tools," http://www.austintexas.gov/department/free-water-conservation-tools.
6. "Irrigation Systems Evaluations," http://www.austintexas.gov/department/irrigation-system-evaluations-and-rebates.
7. "3C Business Challenge: Commit, Calculate, Conserve," http://www.austintexas.gov/department/3c-business-challenge-commit-calculate-conserve.
8. "Free Showerheads and Faucet Aerators Multi-Family Properties," (2019), http://www.austintexas.gov/sites/default/files/files/Water/Conservation/Rebates_and_Programs/Free-Showerheads-Aerators-Multifamily-Facilities.pdf.
9. "Waterwise Hotel Partner Hospitality Partnership," (2017), http://www.austintexas.gov/sites/default/files/files/Water/Conservation/Rebates_and_Programs/WaterWiseHotelPartnership.pdf.
10. "Watering Restrictions," http://www.austintexas.gov/department/watering-restrictions.
11. "Commercial Facility Irrigation Assessment," http://www.austintexas.gov/department/commercial-facility-assessments.
12. "Commercial Vehicle Wash Facility Efficiency Assessment," http://www.austintexas.gov/page/commercial-vehicle-wash-facility-efficiency-assessments.
13. "Cooling Tower Efficiency," http://www.austintexas.gov/page/cooling-towers.
14. "Rebates, Tools, and Programs," http://www.austintexas.gov/department/water-conservation-rebates.
15. "Reclaimed Water," http://www.austintexas.gov/page/reclaimed-water.
16. "Onsite Water Reuse Systems," http://www.austintexas.gov/page/onsite-water-reuse-systems.
17. "Hornsby Bend Biosolids Management Plan," (2010), http://www.ci.austin.tx.us/water/downloads/hornsbybendbro.pdf.

18. "Biogas to Power Wastewater Treatment Plant," http://austinenergy.
blogspot.com/2012/08/biogas-to-power-wastewater-treatment.html.
19. "Dillo Dirt," http://www.austintexas.gov/dillodirt.

REFERENCES

Austin Water. "3C Business Challenge: Commit, Calculate, Conserve." http://
www.austintexas.gov/department/3c-business-challenge-commit-calcu-
late-conserve.
———. "2017–2018 Water & Wastewater Rates." (2018). http://
www.austintexas.gov/sites/default/files/files/Water/Rates/
ResidentialPublicRates_05.2018.pdf.
———. "Biogas to Power Wastewater Treatment Plant." http://austinenergy.
blogspot.com/2012/08/biogas-to-power-wastewater-treatment.html.
———. "Commercial Facility Irrigation Assessment." http://www.austintexas.
gov/department/commercial-facility-assessments.
———. "Commercial Vehicle Wash Facility Efficiency Assessment." http://
www.austintexas.gov/page/commercial-vehicle-wash-facility-efficiency-
assessments.
———. "Cooling Tower Efficiency." http://www.austintexas.gov/page/
cooling-towers.
———. "Dillo Dirt." http://www.austintexas.gov/dillodirt.
———. "Dropcountr Home Water Use Report Frequently Asked Questions."
(2019, January 9). http://www.austintexas.gov/sites/default/files/files/
Water/Conservation/DropcountrFAQ.pdf.
———. "Free Showerheads and Faucet Aerators Multi-Family Properties."
(2019). http://www.austintexas.gov/sites/default/files/files/Water/
Conservation/Rebates_and_Programs/Free-Showerheads-Aerators-
Multifamily-Facilities.pdf.
———. "Free Water Conservation Tools." http://www.austintexas.gov/
department/free-water-conservation tools.
———. "Hornsby Bend Biosolids Management Plan." (2010). http://www.
ci.austin.tx.us/water/downloads/hornsbybendbro.pdf.
———. "Irrigation Systems Evaluations." http://www.austintexas.gov/
department/irrigation-system-evaluations-and-rebates.
———. "Onsite Water Reuse Systems." http://www.austintexas.gov/page/
onsite-water-reuse-systems.
———. "Rebates, Tools, and Programs." http://www.austintexas.gov/
department/water-conservation-rebates.
———. "Reclaimed Water." http://www.austintexas.gov/page/reclaimed-water.
———. "A Water Plan for the Next 100 Years." (2018). http://austintexas.
gov/sites/default/files/files/Water/WaterForward/Water_Forward_
Booklet.pdf.

———. "Watering Restrictions." http://www.austintexas.gov/department/watering-restrictions.

———. "WaterWise Hotel Partner Hospitality Partnership." (2017). http://www.austintexas.gov/sites/default/files/files/Water/Conservation/Rebates_and_Programs/WaterWiseHotelPartnership.pdf.

Office of Sustainability, City of Austin. "Climate Change." https://www.austintexas.gov/climate.

CHAPTER 10

New York City Department of Environmental Protection Developing the Circular Water Economy

Abstract New York City Department of Environmental Protection has implemented a range of technology policies to facilitate the development of the circular water economy that not only mitigates greenhouse gas emissions but also enhances resilience to climate change.

Keywords Water conservation · Water efficiency · Water reuse · Water recycling · Resource recovery · Renewable energy

INTRODUCTION

The New York City Department of Environmental Protection (DEP) operates a network of 19 reservoirs and three controlled lakes that cover around 2000 square miles of watershed land as far as 125 miles upstate. New York City's drinking water system is the largest unfiltered water supply in the world, delivering approximately one billion gallons of high-quality drinking water each day to nine million New Yorkers. New York City's water supply is at risk from climatic extremes, with the severity of widespread summer drought projected to more than double by 2050. Heavy downpours have been increasing with New York State experiencing a 71% increase from 1958 to 2012 in the amount of precipitation falling in very heavy events. To mitigate emissions, New York City has set the goal of reducing greenhouse gas emissions by 80% by 2050.[1,2]

© The Author(s) 2020   135
R. C. Brears, *Developing the Circular Water Economy*,
Palgrave Studies in Climate Resilient Societies,
https://doi.org/10.1007/978-3-030-32575-6_10

This chapter will discuss how DEP has implemented a range of 3R technology policies (reduce, reuse and recycle, and recover) that facilitate the development of the circular water economy that not only mitigates greenhouse gas emissions but also enhances resilience to climate change.

Developing the Circular Water Economy: Reduce

DEP has implemented a range of technology policies to enhance water conservation and promote water efficiency measures in the development of the circular water economy.

Water Rates

New York City properties are assessed for water and sewer services based on the amount of water consumed between the prior and current meter readings. Water meters are read in cubic feet (one cubic feet=7.48 gallons), with customers billed in hundred cubic foot (HCF) units 1 HCF=748 gallons. Properties are billed at how many HCF units are consumed at the current water rate. The New York City Water Board is responsible for establishing water and sewer rates within the city. As of July 1, 2019, for properties billed by meter, the water charge is $3.99 per 100 cubic feet and the combined charge for water and sewer is $10.33 per 100 cubic feet. The minimum water and sewer charge per metered household is $1.27 per day.[3]

Multi-family Conservation Program

The Multi-Family Conservation Program (MCP) is a per-unit flat-rate billing programme designed for buildings with four or more residential units that are equipped with Automated Meter Reading (AMR) devices and have completed by December 31, 2018 a set of water efficiency measures, if not the annual MCP bill will include a 10% surcharge beginning on July 1, 2019. The water efficiency measures include:

- Having installed water-saving toilets (1.6 gallons) and showerhead/faucets (2.5 gallons per minute) in 70% of all units
- Fixing all toilet leaks, and all other leaks in excess of 50 gallons per day and any leaks between 10 and 250 gallons per day that cost less than $250 to repair
- Upgrading laundry room clothes washers to Energy Star-rated products.[4]

Toilet Replacement Program

The Toilet Replacement Program is a voucher-based programme that provides a $125 discount to customers on the cost of new toilet fixtures. The goal of the programme is to incentivise owners of residential and multi-family buildings on the MCP water billing rate to replace inefficient toilets with high-efficiency WaterSense® certified models.[5]

Automated Meter Reading System

DEP is automating its water meter reading capabilities with an AMR system, which consists of small, low-power radio transmitters connected to individual water meters sending daily readings to a network of rooftop receivers throughout the city. Across the city, all 834,000 DEP customers will have an AMR meter installed, with the whole installation taking three years. The new AMR system sends readings to a computerised billing system up to four times a day and will largely eliminate the need for estimated bills.[6]

My DEP Account

AMR-installed customers can log into their My DEP Account online or via the mobile app on Apple and Android mobile devices to view and manage their consumption on a daily, weekly, monthly, and yearly basis. Customers can also sign up for leak notifications.

Alexa, What's My Water Usage?

DEP customers can now use any Amazon Alexa-enabled device to track their water usage, account balance, payments, and bills with NYC DEP skill. Customers can ask Alexa with the NYC DEP skill questions, such as:

- *Water usage*: Alexa, ask NYC DEP what was my water usage three months ago?
- *Account balance*: Alexa, ask NYC DEP what is my balance?
- *Bills*: Alexa, ask NYC DEP what was my bill last month?

Municipal Water Efficiency Program

The Municipal Water Efficiency Program provides funding for water demand reduction projects in City-owned facilities. Under the programme, DEP has identified opportunities for water efficiency projects and water savings in more than 2000 city properties, with estimated savings of over 9 million gallons of water per day. DEP has established interagency partnerships with the Department of Education, the Department of Parks and Recreation, the Fire Department of New York, the City University of New York, and the New York City Housing Authority to plan and implement water efficiency retrofit projects in schools, parks, playgrounds, recreation centres, public universities, firehouses, and public housing developments.[7]

New York City Water Challenges

The New York City Water Challenges are voluntary challenges designed to encourage non-residential groups to match the 5% city-wide water consumption reduction goal. Participants are asked to calculate baseline water consumption, track water usage in their facilities for 12–24 months, develop water conservation plans, and attend meetings with DEP to discuss their progress. Challenge participants receive formal recognition from DEP for their efforts. In the past, DEP has completed the Water Challenge to Hotels (2014), the Water Challenge to Restaurants (2015), and the Water Challenge to Hospitals (2018).

New York City Water Challenge to Universities
With New York City being home to more university students than any other city in the United States, DEP has issued the New York City Water Challenge to Universities to engage campus faculty staff in water conservation strategies and foster a water conservation ethic among students. For the Challenge, which will last two years and conclude on August 2020, participating universities will set the goal to voluntarily reduce their water usage by 5% from their baseline year. To achieve this goal, DEP's water challenge team will connect campus facility staff with a diverse forum of peers, industry experts, and sustainability organisations that provide contacts, best practices, and technical assistance needed to realise significant reductions in water use and costs. DEP will also host workshops to help the universities identify appropriate strategies for water conservation, including water auditing, upgrading domestic use water fixtures, and smart metering. In addition to fixture retrofits, the universities will also be initiating student water conservation campaigns.[8]

Leak Notification Program

DEP's Leak Notification Program is a new initiative that allows DEP to proactively alert customers to potential water leaks on their property. The programme gives customers the opportunity to sign up online and receive email notifications when their water use increases significantly over a period of several days. Specifically, customers enrolled in the programme will receive an email from DEP if their water usage triples for five consecutive days. In order to enrol in the programme, customers must own a tax class 1 property (one-, two-, or three-family home) or tax class 2 property (four-family homes and larger). The property must also have an AMR device installed.[9]

DEVELOPING THE CIRCULAR WATER ECONOMY: REUSE AND RECYCLE

DEP has implemented a range of technology policies to increase the reuse and recycling of water resources in the development of the circular water economy.

NYC DEP Rain Barrel Program

The NYC DEP Rain Barrel Program is part of New York City's Green Infrastructure Plan that aims to capture stormwater before it can enter the sewer system and therefore reduce combined sewer overflows into local waterways. Since 2008, DEP has given away 60-gallon rain barrels that connect directly to a property owner's downspout to capture and store the stormwater that falls on the rooftop. The collected rainwater can be used to water lawns and gardens or for other non-potable uses.[10]

On-Site Water Reuse Grant Program

The On-Site Water Reuse Grant Program provides commercial, mixed-used, and multi-family residential property owners an incentive to install water reuse systems. The grant promotes the construction of systems that safely use rainwater, blackwater, or greywater at both the building-scale and district-scale by covering a proportion of efficiency technology capital costs. The grants available are summarised in Table 10.1.

Table 10.1 On-site water reuse grants

Grant size	Eligibility criteria	Target water savings	Funding amount
Building-scale	Usually 100,000 square feet or more of residential or commercial space	32,000 gallons per day	Up to $250,000
District-scale	Usually includes the sharing of water between two or more parcels	94,000 gallons per day	Up to $500,000

Comprehensive Water Reuse Program

Buildings with successful on-site water reuse systems are eligible for the Comprehensive Water Reuse Program (CWRP). The CWRP provides a 25% water and wastewater fee discount to customers who install water reuse systems that reduce the building's water consumption by at least 25%.[11]

DEVELOPING THE CIRCULAR WATER ECONOMY: RECOVER

DEP has implemented a range of technology policies to recover resources in the development of the circular water economy.

Food Waste and Sludge into Biogas

Food waste in New York City is being collected and sent to Waste Management's CORe® facility in Brooklyn where it is processed into EBS®, which is an organic slurry product used to generate green energy. EBS® is then transported two miles to Newtown Creek Wastewater Treatment Plant where it is mixed together with sludge and then pumped into the digester eggs. Work is underway to purify the biogas generated onsite and turn it into pipeline-quality renewable natural gas for direct in-home use. Currently, Newtown Creek Wastewater Treatment Plant digests around 25,000 gallons of EBS® per day, in addition to over 650,000 gallons of liquid sludge. With the typical U.S. household generating around 3–5 pounds of food waste per day, 25,000 gallons of EBS® is equal to about 100 tonnes of food waste. By mid-2019, the aim is to co-digest 250 tonnes of food scraps per day.[12]

Rooftop Solar Array on the Wastewater Treatment Plant

In 2005, DEP invested $30 million to reduce greenhouse gas emissions from the Port Richmond Wastewater Treatment Plant. New boilers, exhaust system, and one of the city's largest rooftop solar arrays reduce emissions from plant operations by more than 25,000 metric tonnes, the equivalent of removing nearly 6000 cars from the road. The solar array generates around 1.6 million kilowatt-hours, supplying up to 10% of the plant's needs. Prior to the installation of the solar panels, the plant's roof was renovated into a cool roof that reduces the amount of energy absorbed by the roof, reflecting more light back to the solar panels, increasing their output. The solar output was built through a public-private partnership with no upfront capital cost to the City, with Con Edison Solutions owning and maintaining the solar array and the City purchasing the electricity.[13]

CASE STUDY SUMMARY

DEP has implemented a range of 3R technology policies that facilitate the development of the circular water economy that not only mitigates greenhouse gas emissions but also enhances resilience to climate change (Table 10.2).

Table 10.2 NYC DEP case study summary

3R principle	Policy innovation	Description
Reduce	Water rates	New York City properties are assessed for water and sewer services based on the amount of water consumed between the prior and current meter readings
	Multi-family Conservation Program	A per-unit flat-rate billing programme designed for buildings with four+ residential units that are equipped with Automatic Meter Reading devices and have completed water efficiency measures
	Toilet Replacement Program	A voucher-based programme that provides a discount to customers on the cost of new toilet fixtures
	Automated Meter Reading System	DEP is automating its water meter reading capabilities with an Automated Meter Reading system. Customers can log into their My DEP Account to view and manage their consumption

(continued)

Table 10.2   (continued)

3R principle	Policy innovation	Description
	Alexa, what's my water usage?	DEP customers can use any Amazon Alexa-enabled device to track their water usage, account balance, payments, and bills
	Municipal Water Efficiency Program	Provides funding for water demand reduction projects in City-owned facilities
	New York City Water Challenges	Voluntary challenges designed to encourage non-residential groups to match the 5% city-wide water consumption reduction goal. DEP has issued the New York City Water Challenge to Universities to implement water conservation strategies and foster a water conservation ethic among students
	Leak Notification Program	DEP proactively alerts signed-up customers to potential water leaks on their property
Reuse and recycle	NYC DEP Rain Barrel Program	Since 2008, DEP has given away rain barrels. The collected rainwater can be used to water lawns and gardens or for other non-potable uses
	On-Site Water Reuse Grant Program	The grant promotes the construction of rainwater, blackwater, or greywater systems at both the building-scale and district-scale. Customers who install water reuse systems that reduce the building's water consumption by at least 25% receive a 25% fee discount
Recover	Food Waste and Sludge into Biogas	Work is underway to purify the biogas generated at Newtown Creek Wastewater Treatment Plant into pipeline-quality renewable natural gas for direct in-home use
	Rooftop Solar Array on the Wastewater Treatment Plant	Port Richmond Wastewater Treatment Plant has one of New York City's largest rooftop solar arrays with it built through a public-private partnership with no upfront capital cost to the City, with Con Edison Solutions owning and maintaining the solar array and the City purchasing the electricity

Notes

1. The City of New York, "One New York: The Plan for a Strong and Just City," (2015), https://onenyc.cityofnewyork.us/wp-content/uploads/2018/04/OneNYC-1.pdf.
2. States at Risk, "New York," http://statesatrisk.org/new-york/all.

3. NYC DEP, "How We Bill You," https://www1.nyc.gov/site/dep/pay-my-bills/how-we-bill-you.page#targetText=Water%20Rates&target Text=As%20of%20July%201%2C%202019,remains%20at.%20 %241.27%20per%20day.
4. "Multi-Family Conservation Program (Mcp) Frequently Asked Questions," (2018), http://www.nyc.gov/html/dep/pdf/mcpfaq.pdf.
5. "Toilet Replacement Program," https://www1.nyc.gov/html/dep/html/ways_to_save_water/residential-water-efficiency.shtml.
6. "About Automated Meter Reading (Amr)," https://www1.nyc.gov/html/dep/html/customer_services/amr_about.shtml.
7. "Municipal Water Efficiency Program," https://www1.nyc.gov/html/dep/html/ways_to_save_water/municipal-water-efficiency.shtml.
8. "New York City Water Challenge to Universities," https://www1.nyc.gov/html/dep/html/ways_to_save_water/nyc-water-challenge.shtml.
9. "Leak Notification—Sign up Today," https://www1.nyc.gov/html/dep/html/water_and_sewer_bills/leak_notification.shtml.
10. "About the Nyc Dep Rain Barrel Program," https://www1.nyc.gov/html/dep/html/stormwater/rainbarrel.shtml.
11. "On-Site Water Reuse Grant Program," https://www1.nyc.gov/html/dep/html/ways_to_save_water/on-site-water-reuse-grant-program.shtml.
12. New York Water, "Closing the Loop: When Wastewater Treatment Becomes Resource Recovery," https://medium.com/nycwater/closing-the-loop-when-wastewater-treatment-becomes-resource-recovery-8266d1d576cc.
13. NYC DEP, "Department of Environmental Protection Invests $30 Million to Significantly Reduce Greenhouse Gas Emissions from the Port Richmond Wastewater Treatment Plant," https://www1.nyc.gov/html/dep/html/press_releases/15-085pr.shtml#.XGEfFVwzY2x.

References

New York Water. "Closing the Loop: When Wastewater Treatment Becomes Resource Recovery." https://medium.com/nycwater/closing-the-loop-when-wastewater-treatment-becomes-resource-recovery-8266d1d576cc.

NYC DEP. "About Automated Meter Reading (Amr)." https://www1.nyc.gov/html/dep/html/customer_services/amr_about.shtml.

———. "About the Nyc Dep Rain Barrel Program." https://www1.nyc.gov/html/dep/html/stormwater/rainbarrel.shtml.

———. "Department of Environmental Protection Invests $30 Million to Significantly Reduce Greenhouse Gas Emissions from the Port Richmond

Wastewater Treatment Plant." https://www1.nyc.gov/html/dep/html/press_releases/15-085pr.shtml#.XGEfFVwzY2x.

———. "How We Bill You." https://www1.nyc.gov/site/dep/pay-my-bills/how-we-bill-you.page#targetText=Water%20Rates&targetText=As%20of%20July%201%2C%202019,remains%20at%20%241.27%20per%20day.

———. "Leak Notification—Sign up Today." https://www1.nyc.gov/html/dep/html/water_and_sewer_bills/leak_notification.shtml.

———. "Multi-Family Conservation Program (Mcp) Frequently Asked Questions." (2018). http://www.nyc.gov/html/dep/pdf/mcpfaq.pdf.

———. "Municipal Water Efficiency Program." https://www1.nyc.gov/html/dep/html/ways_to_save_water/municipal-water-efficiency.shtml.

———. "New York City Water Challenge to Universities." https://www1.nyc.gov/html/dep/html/ways_to_save_water/nyc-water-challenge.shtml.

———. "On-Site Water Reuse Grant Program." https://www1.nyc.gov/html/dep/html/ways_to_save_water/on-site-water-reuse-grant-program.shtml.

———. "Toilet Replacement Program." https://www1.nyc.gov/html/dep/html/ways_to_save_water/residential-water-efficiency.shtml.

States at Risk. "New York." http://statesatrisk.org/new-york/all.

The City of New York. "One New York: The Plan for a Strong and Just City." (2015). https://onenyc.cityofnewyork.us/wp-content/uploads/2018/04/OneNYC-1.pdf.

CHAPTER 11

South Australia Water Corporation Developing the Circular Water Economy

Abstract SA Water has implemented a range of technology policies to facilitate the development of the circular water economy that not only mitigates greenhouse gas emissions but also enhances resilience to climate change.

Keywords Water conservation · Water efficiency · Water reuse · Water recycling · Resource recovery · Renewable energy

INTRODUCTION

South Australia Water Corporation (SA Water) is owned by the South Australian Government and provides water services to more than 1.7 million South Australian customers. Most of the water supplied is sourced from a range of different places to supply South Australia. Most of the water comes from the River Murray, but surface water, sea water, and groundwater are also used in the supplies. South Australia is the driest state in the driest inhabited continent on Earth and therefore the priority is a secure supply of high-quality water in a variable climate.[1] Regarding greenhouse gas emissions, SA Water has set its reduction target of emissions being no greater than 40% of 1990s greenhouse gas emissions by 31 December 2050.[2] This chapter will discuss how SA Water has implemented a range of 3R technology policies (reduce, reuse and recycle, and recover) that facilitate the development of the circular water economy that not only mitigates greenhouse gas emissions but also enhances resilience to climate change.

DEVELOPING THE CIRCULAR WATER ECONOMY: REDUCE

SA Water has implemented a range of technology policies to enhance water conservation and promote water efficiency measures in the development of the circular water economy.

Residential Water Prices

There is a state-wide price for most water services, with customers in metropolitan or regional areas paying the same price per kilolitre (kL), no matter how much it costs to provide that water. The water charge comprises two parts: a fixed charge for water supply and a variable charge for water use (Table 11.1). The price for recycled water is set at $2.147/kL in 2019–2020, which is 90% of tier 1 water use prices.

Commercial Water Prices

The 2019–2020 water supply charge for commercial customers is based on the property value with a minimum charge of $75.40 per quarter and the water use price of $3.413/kL (or $0.003413 per litre).[3]

Customer Services Online

To meet public expectations of responsive service, accessible information, and more opportunities to interact, SA Water launched *my*SAWater, a

Table 11.1 Residential water use prices

Tier	Usage charge	Price per litre	Indicative quarterly threshold	Daily threshold
1	$2.392/kL	$0.002392	0–30 kL	0–0.3288 kL
2	$3.413/kL	$0.003413	30–130 kL	0.3288 kL–1.4247 kL
3	$3.699/kL	$0.003699	Above 130 kL	Above 1.4247 kL

new online portal that offers customers a variety of services including the ability to receive eBills, pay their bills online, and view their water use, with comparison data available.[4]

Penneshaw Smart Meters Pilot Program

In August 2018, SA Water began its 12-month smart meter pilot programme in Penneshaw with flow and pressure sensors installed along with 300 smart meters for residential and business customers, both of which can receive their smart meter data via the customer portal *myS-martWater*. The programme expands the installation of smart meters beyond the Adelaide Central Business District (CBD) to Penneshaw, Kangaroo Island, enabling the better monitoring and managing of safe clean water. In the CDB, the smart water network incorporates 300 sensors–water flow, pressure, water quality plus acoustic leak detection–along with smart meters for 70 business customers, large and small, to help them better manage their water use and costs. Overall, this combination of technology enables SA Water to identify leaks and fix them overnight, minimising service interruption and commuter delays.[5]

Customer Water Use Portal

SA Water's smart metering system includes a Customer Water Use Portal to help non-residential customers manage their water consumption more effectively. Every 15 minutes, data is collected and forwarded to the portal daily, enabling customers to:

- Increase their understanding of water consumption, when and where water is being used
- View and create water consumption reports on daily, weekly or monthly consumption
- Set email notification alarms for potential leaks, trends, and inconsistencies
- Monitor different parts of the business such as plant equipment, production areas or tenants
- Improve efficiency of environmental monitoring and reporting
- Centralise water monitoring for those with multiple properties.[6]

Leak Analysis and Water Use Profiling Service

SA Water's Business Relations team can help business customers find the best ways to improve water efficiency and find leaks. The service can be tailored to include:

- A site walk-through
- Installation of data loggers on the water meters to track water use for four to six weeks
- Analysis of the collected data
- Quantification of potential leaks
- A written report
- A copy of the raw data.[7]

Educational Resources

SA Water produces a range of educational resources for use in classrooms. The utility also runs the Brainwave Learning Program, which is a series of free events and learning resources that supports the Australian Curriculum in geography, science, and sustainability (Table 11.2).[8,9]

Smartphone Apps

SA Water provides a variety of smartphone apps to educate young people on water conservation and efficiency including the following:

Operation Aqua
This strategic real-time management game puts the player in control of the water and the population's happiness. The game involves the user managing various water resources to ensure demand is satisfied while having to keep costs to a minimum and be aware of any environmental impacts. The game has strong links with the Australian Geography Curriculum, particularly for years 7–10.[10]

Captain Plop's Water-Saving Mission App
SA Water's popular children's education book Captain Plop's Water-Saving Mission is also available as a free app for smartphones and tablets. The book has been written for children from 4 to 8-years old and is

Table 11.2 SA Water's educational resources and learning programmes

Education resources	Programme	Description
Educational tools	Captain Plop books	The Captain Plop books are a series of three titles that deal with water conservation, desalination, and sewage recycling
	Chatterbox	Children create foldable paper toys with each toy having a number of water messages printed on it
	Videos	A range of educational videos covering topics including desalination, sewage treatment, and water testing and treatment are available on SA Water's YouTube channel
Brainwave learning program	Living loans incursion program	Borrow a SA Water staff member for your classroom and discover a wealth of knowledge about water and wastewater treatment, desalination, water cycle and conservation, and the River Murray and salinity issues
	Water treatment: Follow that drop!	Join that science gang for the story of a drop of water's journey to your tap. Classes can tour a water treatment plant and/or participate in hands-on workshops
	Sustainability challenge	This programme challenges students to explore urban water supply and sustainability issues. Students work in teams to design a sustainable town with a water supply that meets growing demands on the supply system. The programme introduces students to water sources, treatment processes, the water cycle, urban planning, populations, and demographics

designed to help them understand the importance of water conservation around the home.[11]

Developing the Circular Water Economy: Reuse and Recycle

SA Water has implemented a range of technology policies to increase the reuse and recycling of water resources in the development of the circular water economy.

Greywater Regulated

Customers that wish to install greywater systems must receive approval for each part of the system: treatment, development, and installation. Each approval comes from a different place, including SA Health and the local government, to ensure that the overall system will protect both human health and the environment.[12]

Recycled Water

Recycled water is supplied on a separate system using purple lines. It has been treated to a standard that is safe for a range of household purposes including flushing toilets, watering gardens, and washing cars. SA Water supplies recycled water to:

- *Adelaide Park Lands*: Recycled water from the Glenelg Wastewater Treatment Plant is used to keep the Adelaide Park Lands green during the year with a minimum of 1.3 billion litres supplied annually
- *Adelaide's southern suburbs*: Sewage collected at the Christies Beach Wastewater Treatment Plant is treated and supplied to around 8000 homes in the southern suburbs
- *Virginia's market gardeners*: The Virginia Pipeline Scheme uses recycled water from the Bolivar Wastewater Treatment Plant to supply around 360 customers in the Virginia area north of Adelaide.[13]

Northern Adelaide Irrigation Scheme

The Northern Adelaide Irrigation Scheme (NAIS) will see new water treatment facilities built with the Bolivar precinct to increase its production of recycled irrigation water by 60%. NAIS will deliver up to 12 gigalitres (GL) of reclaimed water suitable for commercial food production. NAIS infrastructure will treat, store, and transport climate and season independent water to the farm gate. It is planned that up to 20 GL will be delivered to enable future growth.[14]

Developing the Circular Water Economy: Recover

SA Water has implemented a range of technology policies to recover resources in the development of the circular water economy.

SA Water's Soil Conditioner to South Australian Farmers

Over 30,000 equivalent dry tonnes of biosolids are produced annually in SA Water's wastewater treatment plants. For more than 15 years, South Australian farmers and horticulturalists have been successfully applying biosolids to thousands of hectares of dryland agriculture to grow cereal crops, including wheat, barley, and canola, and irrigated permanent plantings such as vines and olives. SA Water assists farmers to use biosolids on their properties and has been arranging soil surveys, analyses, and necessary paperwork required for approval for use from the SA Environmental Protection Agency. Biosolids are available free of charge from selected SA Water wastewater treatment plants.[15]

Thermal Energy Storage

SA Water is trialling new silicon thermal energy storage technology at its Glenelg Wastewater Treatment Plant, which is already 80% energy self-sufficient from burning biogas to generate electricity through reciprocating gas engines that are then used in the facility. By integrating storage, it would enable the plant to avoid having to use biogas as its produced and instead use it to coincide with peak operational patterns and high electricity market prices. The technology will store latent heat in molten silicon at 1414 degrees Celsius to form a 10 MWh thermal storage device. This will release around 250 kW for six hours as well as heat which is returned to the plant's digesters.[16]

Solar Photovoltaic and Battery Storage Trial

SA Water will be trialling a solar photovoltaic (PV) and battery storage system at its Crystal Brook workshop to increase its energy

Table 11.3 SA Water case study summary

3R principle	Policy innovation	Description
Reduce	Residential Water Prices	There is a state-wide price for most water services, with customers in metropolitan or regional areas paying the same price per kilolitre, no matter how much it costs to provide that water
	Commercial Water Prices	The water supply charge for commercial customers is based on the property value with a minimum charge and a per kL consumption charge
	Customer Services Online	mySAWater is a new online portal that lets customers view their water use, with comparison data available
	Penneshaw Smart Meters Pilot Program	SA Water has begun its 12-month smart meter pilot programme in Penneshaw with flow and pressure sensors installed along with 300 smart meters for residential and business customers
	Leak Analysis and Water Use Profiling Service	SA Water's business relations team can help business customers find the best ways to improve water efficiency and find leaks
	Educational Resources	SA Water produces a range of educational resources for use in classrooms including books, toys, and videos
	Brainwave Learning Program	A series of free events and learning resources that supports the Australian Curriculum in the areas of geography, science, and sustainability
	Smartphone Apps	Operation aqua This strategic real-time management game puts the player in control of the water and the population's happiness
		Captain Plop's Water-Saving Mission App SA Water's popular children's education book Captain Plop's Water-Saving Mission is also available as a free app
Reuse and recycle	Greywater Regulated	Greywater systems must receive approval for each part of the system from a range of agencies including SA Health and the local government, to ensure the protection of human health and the environment
	Recycled Water	Recycled water is supplied on a separate system using purple lines. It has been treated to a standard that is safe for a range of household purposes
	Northern Adelaide Irrigation Scheme	The scheme will see new water treatment facilities built with the Bolivar precinct to increase its production of recycled irrigation water by 60%. The scheme will deliver climate and season independent water to the farm gate
Recover	SA Water's Soil Conditioner to South Australian Farmers	Biosolids are available free of charge from selected SA Water wastewater treatment plants
	Thermal Energy Storage	SA Water is trialling new silicon thermal energy storage technology at its Glenelg Wastewater Treatment Plant
	Solar Photovoltaic and Battery Storage Trial	SA Water will be trialling a solar photovoltaic and battery storage system at its Crystal Brook workshop
	Floating Solar Panels	SA Water will trial floating solar panel arrays on its Happy Valley Reservoir with the installation of a 100-kilowatt pilot system

self-sufficiency and contribute to the corporation's goal of achieving zero net electricity costs by 2020. The installation of 100 kW of solar PV and 50 kWh of battery storage will enable the utility to validate the effectiveness of the technology in reducing costs.[17]

Floating Solar Panels

SA Water will trial floating solar panel arrays on its Happy Valley Reservoir with the installation of a 100-kilowatt pilot system. In addition to producing electricity to power the nearby Happy Valley Water Treatment Plant, the floating solar panels may also reduce evaporation and the incidence of algal blooms.[18]

CASE STUDY SUMMARY

SA Water has implemented a range of 3R technology policies that facilitate the development of the circular water economy that not only mitigates greenhouse gas emissions but also enhances resilience to climate change (Table 11.3).

NOTES

1. SA Water, "2017–18 South Australian Water Corporation Annual Report," (2018), https://www.sawater.com.au/__data/assets/pdf_file/0009/310887/2017-18-Annual-Report-accessible.pdf.
2. "Environmental Commitments," https://www.sawater.com.au/community-and-environment/environmental-commitments.
3. "Water & Sewerage Prices," https://www.sawater.com.au/accounts-and-billing/current-water-and-sewerage-rates.
4. "2017–18 South Australian Water Corporation Annual Report".
5. "Penneshaw Smart Meters Pilot Program," https://www.sawater.com.au/residential/penneshaw-smart-meters-pilot-program.
6. "Customer Water Use Portal," https://www.sawater.com.au/business/products-and-services/customer-water-use-portal2.
7. "Leak Analysis and Water Use Profiling Service," https://www.sawater.com.au/business/products-and-services/leak-analysis-and-water-use-profiling-service.
8. "Education Resources," https://www.sawater.com.au/community-and-environment/schools/education-resources.

9. "Brainwave Learning Program," https://www.sawater.com.au/community-and-environment/schools/brainwave-learning-program.
10. "Apps and Games," https://www.sawater.com.au/community-and-environment/schools/online-learning/apps-and-games.
11. Ibid.
12. "Greywater," https://www.sawater.com.au/residential/water-in-your-home-and-garden/greywater.
13. "Recycled Water," https://www.sawater.com.au/community-and-environment/our-water-and-sewerage-systems/recycled-water.
14. "Northern Adelaide Irrigation Scheme (Nais)," https://www.sawater.com.au/current-projects/nais.
15. "SA Water's Soil Conditioner to South Australian Farmers," (2015), https://www.livestocksa.org.au/media/documents/SA%20Water%20Fact%20Sheet%20-%20Biosolids%20Apr2015%20V1%201.pdf.
16. "Poo Puts Downward Pressure on Power Costs," https://www.sawater.com.au/news/poo-puts-downward-pressure-on-power-costs.
17. "Crystal Brook Lights up SA Water's Energy Future," https://www.sawater.com.au/news/crystal-brook-lights-up-sa-waters-energy-future.
18. "Solar Floats Electricity Costs to Zero," https://www.sawater.com.au/news/solar-floats-electricity-costs-to-zero.

References

SA Water. "2017–18 South Australian Water Corporation Annual Report." (2018). https://www.sawater.com.au/__data/assets/pdf_file/0009/310887/2017-18-Annual-Report-accessible.pdf.

———. "Apps and Games." https://www.sawater.com.au/community-and-environment/schools/online-learning/apps-and-games.

———. "Brainwave Learning Program." https://www.sawater.com.au/community-and-environment/schools/brainwave-learning-program.

———. "Crystal Brook Lights up SA Water's Energy Future." https://www.sawater.com.au/news/crystal-brook-lights-up-sa-waters-energy-future.

———. "Customer Water Use Portal." https://www.sawater.com.au/business/products-and-services/customer-water-use-portal2.

———. "Education Resources." https://www.sawater.com.au/community-and-environment/schools/education-resources.

———. "Environmental Commitments." https://www.sawater.com.au/community-and-environment/environmental-commitments.

———. "Greywater." https://www.sawater.com.au/residential/water-in-your-home-and-garden/greywater.

———. "Leak Analysis and Water Use Profiling Service." https://www.sawater.com.au/business/products-and-services/leak-analysis-and-water-use-profiling-service.

———. "Northern Adelaide Irrigation Scheme (Nais)." https://www.sawater.com.au/current-projects/nais.

———. "Penneshaw Smart Meters Pilot Program." https://www.sawater.com.au/residential/penneshaw-smart-meters-pilot-program.

———. "Poo Puts Downward Pressure on Power Costs." https://www.sawater.com.au/news/poo-puts-downward-pressure-on-power-costs.

———. "Recycled Water." https://www.sawater.com.au/community-and-environment/our-water-and-sewerage-systems/recycled-water.

———. "SA Water's Soil Conditioner to South Australian Farmers." (2015). https://www.livestocksa.org.au/media/documents/SA%20Water%20Fact%20Sheet%20-%20Biosolids%20Apr2015%20V1%201.pdf.

———. "Solar Floats Electricity Costs to Zero." https://www.sawater.com.au/news/solar-floats-electricity-costs-to-zero.

———. "Water & Sewerage Prices." https://www.sawater.com.au/accounts-and-billing/current-water-and-sewerage-rates.

CHAPTER 12

San Francisco Public Utilities Commission Developing the Circular Water Economy

Abstract San Francisco Public Utilities Commission has implemented a range of technology policies to facilitate the development of the circular water economy that not only mitigates greenhouse gas emissions but also enhances resilience to climate change.

Keywords Water conservation · Water efficiency · Water reuse · Water recycling · Resource recovery · Renewable energy

INTRODUCTION

San Francisco Public Utilities Commission (SFPUC) provides water services to 2.7 million residents and businesses in the Bay Area. The water supply system stretches from the Sierra to the City and features a complex series of reservoirs, tunnels, pipelines, and treatment systems. In addition to the water being among the purest in the world, the system for delivering water is almost gravity-fed, requiring minimal fossil fuel to move and treat the water. A goal of SFPUC is that the City will optimise the use of its finite water and energy resources to create a more resilient and reliable future while at the same time meeting community and ecosystem needs. This chapter will discuss how SFPUC has implemented a range of 3R technology policies (reduce, reuse and recycle, and recover)

R. C. Brears, *Developing the Circular Water Economy*,
Palgrave Studies in Climate Resilient Societies,
https://doi.org/10.1007/978-3-030-32575-6_12

that facilitate the development of the circular water economy that not only mitigates greenhouse gas emissions but also enhances resilience to climate change.

DEVELOPING THE CIRCULAR WATER ECONOMY: REDUCE

SFPUC has implemented a range of technology policies to enhance water conservation and promote water efficiency measures in the development of the circular water economy.

Drought Surcharge

SFPUC's water rate structure consists of a fixed monthly service charge based on meter size and a variable charge which is based on volumetric usage. The variable charge for residential customers is comprised of a two-tier, inclining block rate structure, while non-residential customers are charged a uniform commodity rate (Table 12.1). If SFPUC's Commission adopts a resolution declaring a stage of water delivery reduction, a drought surcharge is applied to water rates, which is listed in Table 12.2.[1]

Table 12.1 Volumetric charge for water

Customer type	First units per month	Additional units
Single-family residential	$7.85 (first four units)	$9.61
Multi-family residential	$7.94 (first three units)	$9.73
Commercial, industrial, and general uses	$9.14 (all units of water)	
(1 unit = 1 Ccf of water = 748 Gallons)		

Table 12.2 Drought surcharge

Retail water shortage allocation plan stage	Target usage reduction	Drought surcharge on volumetric water rates
Stage 1	5–10%	Up to 10%
Stage 2	11–20%	Up to 20%
Stage 3	Over 20%	Up to 25%

Free Evaluations and Devices

SFPUC offers a range of free evaluations and devices to encourage customers to use water wisely including the following:

Water-Wise Evaluations

SFPUC offers free Water-Wise Evaluations to all water account holders, property owners, managers, and at times tenants of properties to help them identify costly leaks and replace inefficient fixtures. Customers also receive tips on how to use indoor and outdoor water more efficiently. Single-family and small multi-family and commercial evaluations take around one-two hours, while larger multi-family and non-residential evaluations typically require a half or full day.[2]

Landscape Technical Assistance Program

For landscapes that are greater than half-acre, SFPUC offers the Landscape Technical Assistance Program (LTAP) that can help retail water users identify water-saving opportunities that can improve irrigation efficiency, reduce water use, and save money. Participants in LTAP will receive a site evaluation conducted by an expert landscape architect and irrigation designer, as well as a customised report including:

- A summary of existing irrigation system operating conditions and assessment of landscape plantings
- Recommended landscape planting modifications for replacing high water use plants with alternative low water use species
- Prioritised recommendations that improve irrigation efficiency and plant health
- An estimate of potential annual water and cost savings
- Cost estimate for implementing recommendations
- Site-specific water budget and maintenance guide.[3]

Grant Assistance for Water Efficient Equipment Retrofits

Grant assistance is available for non-residential customers who can significantly reduce their use of potable water through upgrade or replacement of existing onsite indoor water-using equipment. Eligible projects must achieve a water saving of 200 ccf (149,000 gallons) or more a year to qualify. SFPUC will provide qualifying projects with grant funding of $1.00 per ccf over a 10-year lifespan up to 50% of the project's equipment costs. Qualifying projects include fixed water-saving retrofit projects that consist of standardised equipment with predictable water savings projects or custom retrofit projects that consist of unique or site-specific equipment retrofits that result in project-specific water savings.[4]

DEVELOPING THE CIRCULAR WATER ECONOMY: REUSE AND RECYCLE

SFPUC has implemented a range of technology policies to increase the reuse and recycling of water resources in the development of the circular water economy.

Green Infrastructure Grant Program

SFPUC has launched the Green Infrastructure Grant Program to encourage property owners to design, build, and maintain performance-based green stormwater infrastructure including, but not limited to, rainwater harvesting, permeable pavement, rain gardens, and vegetated roofs. The goal of the programme is to reduce the volume of stormwater runoff entering SFPUC's sewer system and improve system performance while also providing co-benefits including non-potable reuse, groundwater recharge, and workforce development. To receive funding, the applicant must demonstrate that the project is located on land connected to SFPUC's sewer system, manages stormwater runoff from a minimum impervious area of 0.5 acres, captures the 90th percentile (0.75-inch depth) with the proposed green infrastructure features, and provides at least two of the identified co-benefits of:

- Being located within a disadvantaged community
- Provides public access opportunities
- Facilitates groundwater recharge
- Encourages non-potable use
- Provides educational or curriculum opportunities
- Facilities job and training opportunities
- Increases biodiversity/native habitat.[5]

Individual grant awards are capped at a maximum of $765,000 per impervious acre managed (i.e. the amount of impervious surface that drains to the green infrastructure, or 'impervious acres managed'), up to a maximum of $2 million per grant.[6]

Greywater

SFPUC has developed a laundry-to-landscape greywater programme and a rebate to encourage the uptake of greywater systems.

Laundry-to-Landscape Graywater Program

SFPUC's Laundry-to-Landscape (L2L) Graywater Program, in partnership with Urban Farmer Store, offers residents a $125 discount off the purchase of an L2L greywater kit (retail cost $175). The kit includes the basic components to divert clothes washer water to gardens with participants also receiving:

- A copy of the *San Francisco Graywater Design Manual for Outdoor Irrigation*
- Optional on-site consultation with a greywater expert
- Mandatory training to review the design, installation, and maintenance requirements of an L2L greywater system
- Access to a free tool kit for do-it-yourself installation.

Graywater Permit Rebate Program

For greywater projects that require a permit from the Department of Building Inspection (required for all other greywater systems except L2L), SFPUC's Graywater Permit Rebate Program offers a rebate up to $225 to help cover the costs of the permit. The rebate is available for residential greywater systems with subsurface irrigation.[7]

Onsite Water Reuse for Commercial, Multi-family, and Mixed Use Development Ordinance

In 2012, the City and County of San Francisco adopted the Onsite Water Reuse for Commercial, Multi-family, and Mixed Use Development Ordinance. Known as the Non-potable Water Ordinance, it allows for the collection, treatment, and use of alternate water sources for non-potable applications in individual buildings and at the district scale. Since 2015, new development projects of 250,000 square feet or more of gross area that have not received a site permit prior to November 1st, 2016, are required to install and operate an onsite non-potable water system to treat and reuse greywater, rainwater, and foundation drainage for toilet and urinal flushing and irrigation. Meanwhile, new development projects of 40,000 square feet or more of gross floor area are required to prepare water budget calculations that assess the amount of available rainwater, greywater, and foundation drainage and the demands for toilet and urinal flushing and irrigation.

Non-potable Grant Program

The SFPUC Non-potable Grant Program encourages retail water users to collect, treat, and use alternate water sources including rainwater, stormwater, greywater, foundation water, blackwater, and brewery process water for non-potable uses. The programme targets projects that can reduce potable water use by maximising available onsite alternate water sources for toilet flushing, irrigation, and other non-potable uses. Projects eligible for the grant must meet one of the following criteria listed in Table 12.3.[8]

Groundwater Program

SFPUC's Groundwater Program increases local and regional water supply reliability, diversifies the city's water supply portfolio, and reduces vulnerability to disruptions from drought and natural disasters.

Regional Groundwater Storage and Recovery Project
The Regional Water System includes a blend of surface water from Hetch Hetchy Reservoir in Yosemite National Park and five Bay Area reservoirs located in Alameda and San Mateo counties. During periods of drought, these surface water supplies can be severely impacted, affecting the reliability of water supply for the region. The Regional

Table 12.3 Non-potable grant program criteria

Criteria	Grant funding available
Projects that replace at least 450,000 gallons of potable water per year	Up to $100,000
Projects that replace at least 1,000,000 gallons of potable water per year	Up to $250,000
Projects that replace at least 3,000,000 gallons of potable water per year	Up to $500,000

Groundwater Storage and Recovery Project, currently under construction with completion anticipated for 2021, will coordinate the management of surface and groundwater supplies to meet drought year water supply needs for the Regional Water System. Once completed, it will comprise 16 deep groundwater wells and facilities consisting of chemical treatment equipment, tanks, pumping systems, and associated pipelines. When operational, the project will store water in wet years and recover that water for use during dry years. As part of the project, surface water will be used instead of groundwater in wet years, allowing the groundwater to recharge through rainfall and decreased pumping. This will create a savings account of up to 20 billion gallons of groundwater that will be stored in the aquifer. In dry years, when there is less surface water available, the saved water will be pumped from the new groundwater well recovery facilities at a rate of 7.2 million gallons per day. Overall, the protected groundwater supply will help augment the Regional Water System supply over a 7.5-year-long drought.[9]

San Francisco Groundwater Supply Project
SFPUC pumps groundwater from the Westside Groundwater Basin aquifer that extends approximately 270 feet–460 feet below the surface. The groundwater is treated and blended with the utility's regional drinking water supplies before it is delivered to customers. Over the next few years, SFPUC will continue adding groundwater to reach the goal of blending four million gallons a day of treated groundwater with the regional water supplies. To date, four wells have been completed with the remaining two still in construction. The groundwater is treated with chlorine and then delivered to the Sunset and Sutro reservoirs. The blended water is served to more than half of the SFPUC's customers in the city.[10]

Recycled Water

SFPUC and the City of San Francisco have implemented recycled water infrastructure in the development of the circular water economy.

Recycled Water Ordinance
To supplement San Francisco's imported water supplies, the City and County of San Francisco's Recycled Water Ordinance requires property owners to install recycled water systems in new construction,

modification, or remodel projects for authorised applications such as landscape irrigation, toilet/urinal flushing, cooling, and water features. The requirements of the Recycled Water Ordinance apply to properties located within the designated recycled water use areas under the circumstances of:

- New construction or major alterations to a building totalling 40,000 square feet or more
- All subdivisions
- New and existing irrigated areas of 10,000 square feet or more not constructed in conjunction with a development project.[11]

Westside Enhanced Water Recycling Project
On the west side of San Francisco, SFPUC is aiming to save up to two million gallons per day on average of drinking water that is currently being used for non-drinking water purposes such as irrigation and lake-fill. The Westside Enhanced Water Recycling Project will provide recycled water for these non-potable uses to conserve potable supplies for drinking. The project includes the construction of a new recycled water treatment facility that will be located within the limits of SFPUC's Oceanside Wastewater Treatment Plant and almost eight miles of new recycled water pipelines. The project will also include an 840,000-gallon underground reservoir and an above-ground recycled water pump station in Golden Gate Park that will pump recycled water up to Lincoln Park Golf Course, the Presidio Golf Course, and other landscaped areas for irrigation.[12]

Developing the Circular Water Economy: Recover

SFPUC has implemented a range of technology policies to recover resources in the development of the circular water economy.

Combined Heat and Power

SFPUC operates all wastewater treatment plants in San Francisco and handles all biogas generated. Part of the energy that powers SFPUC's Oceanside and Southeast wastewater treatment plants is generated from biogas. The biogas-fuelled cogeneration technology provides both heat and electricity for plant operations. In total, San Francisco generates up to 3.2 megawatts of renewable energy from this process.[13]

Biosolids Digester Facilities Project

Built in 1952, the Southeast Treatment Plant (SEP) is San Francisco's largest wastewater facility and treats 80% of the city's sewage and stormwater flows. With much of its infrastructure over 60 years old and not meeting current seismic standards, the Biosolids Digester Facilities Project will replace and relocate the outdated existing solids treatment facilities with more reliable, efficient, and modern technologies that will help transform SEP into a modern resource recovery facility. With construction expected to be complete in 2025, SEP aims to produce higher quality biosolids, capture and treat odours more effectively, and maximise biogas utilisation and energy recovery to produce heat, steam, and energy.[14]

Public-Private Partnership to Sell Biofertiliser

SFPUC has partnered with Lystek to sell a renewable biosolids-based fertiliser to California farms and ranches. Through its wastewater treatment process, SFPUC extracts solids from the incoming wastewater and uses anaerobic digestion to transform the solids into biosolids. Under the partnership, Lystek manages around 15% of SFPUC's biosolids at its Organic Material Recovery Center, located in Fairfield-Suisen Sewer District. At the Organic Material Recovery Center, Lystek uses its patented Lystek Thermal Hydrolysis Process to convert biosolids into LysteGro, a commercial biofertiliser product rich in carbon and nutrients.[15]

Solar Arrays

SFPUC has installed a variety of solar photovoltaic (PV) installations at several water and wastewater-related sites.

North Point Wet Weather Facility Solar Photovoltaic
At the North Point Wet Weather Facility, two separate building roofs are used to provide supplemental solar PV power. Located near Fisherman's Wharf, the equipment was designed to resist corrosion from the salt and other harsh environmental factors.

Southeast Water Pollution Control Plant Solar Photovoltaic System
A solar PV system has been installed on the roofs of two wastewater treatment buildings at the Southeast Water Pollution Control Plant. The system generates 255 kW of renewable energy, supplementing other electrical power at the site which treats 70 million gallons per day of the city's wastewater.

Sunset Reservoir Solar Array

The Sunset Reservoir Solar Array is San Francisco's largest solar installation. The system consists of nearly 24,000 solar panels and has a generating capacity of up to five megawatts. The electricity generated is delivered to SFPUC under a 25-year Power Purchase Agreement, with the renewable energy used to help power public buses, the San Francisco International Airport, health clinics, and other city services.[16]

Hetch Hetchy Hydroelectric Power

San Francisco's drinking water is harnessed to generate hydroelectric power that meets about 17% of the City's electricity needs. The Hetch Hetchy Regional Power System is composed of three hydroelectric powerhouses: Moccasin Powerhouse, which includes a small, in-line hydroelectric unit, Kirkwood Powerhouse, and Holm Powerhouse. The combined total hydroelectric generating capacity of these facilities is around 385 megawatts. Hydroelectric generation at Moccasin and Kirkwood rely on gravity-driven water flowing downhill from the Hetch Hetchy Reservoir while Holm generates energy from gravity-driven water flowing downhill from Cherry Lake. On average, the Hetch Hetchy Power System generates 1.5 billion kilowatt-hours of hydroelectric each year.[17]

SFGreasecycle

SFPUC's SFGreasecycle programme helps prevent fats, oils, and grease (FOG) damage to the City's sewer system by providing free collection services to restaurants and residents across San Francisco. After removal of impurities (food, scraps, and water) and primary polishing, SFPUC sells the grease by-product for conversion into biodiesel. The biofuel can then be sold to city transit fleets to offset their carbon emissions from fossil fuels.[18]

Case Study Summary

SFPUC has implemented a range of 3R technology policies that facilitate the development of the circular water economy that not only mitigates greenhouse gas emissions but also enhances resilience to climate change (Table 12.4).

Table 12.4   SFPUC case study summary

3R principle	Policy innovation	Description	
Reduce	Drought Surcharge	If SFPUC's Commission adopts a resolution declaring a stage of water deliver reduction, a drought surcharge is applied to water rates	
	Free Evaluations and Devices	Water-Wise Evaluations	Offered to all water account holders, property owners, managers, and at times tenants of properties, to help them identify leaks, replace inefficient fixtures, and provide tips on more efficient indoor and outdoor water use
		Landscape Technical Assistance Program	SFPUC can help retail water users identify water saving opportunities
	Grant Assistance for Water Efficient Equipment Retrofits	Grants are available for non-residential customers who can significantly reduce their use of potable water through upgrades or replacement of existing onsite indoor water-using equipment	
Reuse and recycle	Green Infrastructure Grant Program	Encourages property owners to design, build, and maintain performance-based green stormwater infrastructure including rainwater harvesting systems that provide multiple benefits including non-potable reuse and groundwater recharge	
	Greywater	Laundry-to-Landscape Graywater Program	Offers residents a discount off the purchase of a Laundry-to-Landscape greywater kit that includes the basic components to divert clothes washer water to gardens
		Graywater Permit Rebate Program	For greywater projects that require a permit from the Department of Building Inspection, SFPUC offers a rebate to help cover the costs of the permit
	Onsite Water Reuse for Commercial, Multi-family, and Mixed Use Development Ordinance	Since 2015, new large development projects are required to install and operate an onsite non-potable water system	
	Non-potable Grant Program	Encourages retail water users to collect, treat, and use alternate water sources for non-potable uses	
	Groundwater Program	Regional Groundwater Storage and Recovery Project	Will coordinate the management of surface and groundwater supplies to meet drought year water supply needs for the Regional Water System
		San Francisco Groundwater Supply Project	SFPUC treats and blends groundwater with the utility's regional drinking water supplies before it is delivered to customers
	Recycled Water	Recycled Water Ordinance	Requires property owners to install recycled water systems in new construction, modification, or remodel projects
		Westside Enhanced Water Recycling Project	The project will provide recycled water for non-potable uses to conserve potable supplies for drinking

(continued)

Table 12.4   (continued)

3R principle	Policy innovation	Description
Recover	Combined Heat and Power	Part of the energy that powers SFPUC's Oceanside and Southeast wastewater treatment plants is generated from biogas. The biogas-fuelled cogeneration technology provides both heat and electricity for plant operations
	Biosolids Digester Facilities Project	The Southeast Treatment Plant aims to produce higher quality biosolids and maximise biogas utilisation and energy recovery to produce heat, steam, and energy
	Public-Private Partnership to Sell Biofertiliser	SFPUC has partnered with Lystek to sell a renewable biosolids-based fertiliser to California farms and ranches
	Solar Arrays	North Point wet weather facility solar photovoltaic — Two separate building roofs are used to provide supplemental solar power
		Southeast water pollution control plant solar photovoltaic system — A solar photovoltaic system has been installed on the roofs of two wastewater treatment buildings at the plant
		Sunset Reservoir Solar Array — San Francisco's largest solar installation, with excess renewable energy used to help power public buses, the San Francisco International Airport, health clinics, and other city services
	Hetch Hetchy Hydroelectric Power	San Francisco's drinking water is harnessed to generate hydroelectric power that meets about 17% of the City's electricity needs
	SFGreasecycle	Helps prevent fats, oils, and grease damage to the City's sewer system by providing free collection services to restaurants and residents across San Francisco. SFPUC sells the grease by-product for conversion into biodiesel for the city's transit fleets

NOTES

1. SFPUC, "Water and Sewer Rate Information," https://www.sfwater.org/index.aspx?page=168.
2. "Water-Wise Evaluations," https://sfwater.org/index.aspx?page=138.
3. "Large Landscape Assistance Programs," https://sfwater.org/index.aspx?page=103.
4. "Grant Assistance for Water Efficient Equipment Retrofits," https://sfwater.org/index.aspx?page=512.
5. "Green Infrastructure Grant Program," https://sfwater.org/index.aspx?page=1260.
6. Ibid.
7. "Graywater," https://www.sfwater.org/index.aspx?page=100.

8. "Non-potable Water Program," https://sfwater.org/index.aspx?page= 686.
9. "Regional Groundwater Storage & Recovery," https://sfwater.org/ index.aspx?page=982.
10. "San Francisco Groundwater Supply Project," https://sfwater.org/index. aspx?page=1136.
11. "Recycled Water Use," https://sfwater.org/index.aspx?page=687.
12. "Westside Enhanced Water Recycling Project," https://sfwater.org/ index.aspx?page=144.
13. SF Environment, "Biogas," https://sfenvironment.org/article/biomass-amp-biofuels/biogas.
14. SFPUC, "Biosolids Digester Facilities Project," (2018), https://sfwater. org/modules/showdocument.aspx?documentid=12560.
15. Lystek, "Sfpuc Receives First Payment for Sale of Biofertilizer," https:// lystek.com/sfpuc-receives-first-payment-for-sale-of-biofertilizer/.
16. SFPUC, "Solar Installations," https://sfwater.org/index.aspx?page=403.
17. "About the Power Enterprise," https://sfwater.org/index.aspx?page=391.
18. "Sfgreasecycle," https://sfwater.org/index.aspx?page=465.

References

Lystek. "SFPUC Receives First Payment for Sale of Biofertilizer." https://lystek. com/sfpuc-receives-first-payment-for-sale-of-biofertilizer/.
SF Environment. "Biogas." https://sfenvironment.org/article/biomass-amp-biofuels/biogas.
SFPUC. "About the Power Enterprise." https://sfwater.org/index.aspx? page=391.
———. "Biosolids Digester Facilities Project." (2018). https://sfwater.org/ modules/showdocument.aspx?documentid=12560.
———. "Grant Assistance for Water Efficient Equipment Retrofits." https:/// sfwater.org/index.aspx?page=512.
———. "Graywater." https://www.sfwater.org/index.aspx?page=100.
———. "Green Infrastructure Grant Program." https://sfwater.org/index. aspx?page=1260.
———. "Large Landscape Assistance Programs." https://sfwater.org/index. aspx?page=103.
———. "Non-potable Water Program." https://sfwater.org/index.aspx? page=686.
———. "Recycled Water Use." https://sfwater.org/index.aspx?page=687.
———. "Regional Groundwater Storage & Recovery." https://sfwater.org/ index.aspx?page=982.

———. "San Francisco Groundwater Supply Project." https://sfwater.org/index.aspx?page=1136.

———. "Sfgreasecycle." https://sfwater.org/index.aspx?page=465.

———. "Solar Installations." https://sfwater.org/index.aspx?page=403.

———. "Water-Wise Evaluations." https://sfwater.org/index.aspx?page=138.

———. "Water and Sewer Rate Information." https://www.sfwater.org/index.aspx?page=168.

———. "Westside Enhanced Water Recycling Project." https://sfwater.org/index.aspx?page=144.

CHAPTER 13

Singapore's Public Utilities Board Developing the Circular Water Economy

Abstract Singapore's Public Utilities Board has implemented a range of technology policies to facilitate the development of the circular water economy that not only mitigates greenhouse gas emissions but also enhances resilience to climate change.

Keywords Water conservation · Water efficiency · Water reuse · Water recycling · Resource recovery · Renewable energy

INTRODUCTION

Singapore's Public Utilities Board (PUB) is Singapore's national water authority that is responsible for the collection, production, distribution, and reclamation of water. Singapore has developed a diverse water supply from four water sources known as the Four National Taps: water from local catchments, imported water, high-grade reclaimed water known as NEWater, and desalinated water. Currently, domestic customers consume 45% and the non-domestic sector takes up the rest. By 2060, Singapore's total water demand could almost double due to population and economic growth, with the non-domestic sector accounting for around 70%. By then NEWater and desalination will meet up to 85% of future water demand. In addition to implementing water infrastructure to secure an adequate and affordable supply of water, PUB has implemented a wide

© The Author(s) 2020
R. C. Brears, *Developing the Circular Water Economy*,
Palgrave Studies in Climate Resilient Societies,
https://doi.org/10.1007/978-3-030-32575-6_13

range of water conservation initiatives to encourage prudent water usage.[1] This chapter will discuss how PUB has implemented a range of 3R technology policies (reduce, reuse and recycle, and recover) that facilitate the development of the circular water economy that not only mitigates greenhouse gas emissions but also enhances resilience to climate change.

DEVELOPING THE CIRCULAR WATER ECONOMY: REDUCE

PUB has implemented a range of technology policies to enhance water conservation and promote water efficiency measures in the development of the circular water economy.

Components of the Water Price

PUB has three components to the water price in the monthly bill:

- *Water Tariff*: Covers the costs incurred in the various stages of the water production process (collection of rainwater, treatment of raw water, and distribution of treated potable water to customers). The Water Tariff is charged based on the volume of water consumed.
- *Water Conservation Tax*: The Water Conservation Tax (WCT) encourages water conservation and reflects its scarcity value. WCT is imposed as a percentage of the Water Tariff to reinforce that water is precious from the first drop.
- *Waterborne Fee*: The Waterborne Fee (WBF) goes toward the cost of treating used water and maintaining the used water network. It is based on the volume of water usage.

Water Price Revision

PUB has revised the water price for all customers in two phases: The first phase took place on 1 July 2017 and the second revision on 1 July 2018 (Tables 13.1, 13.2 and 13.3). Large investments, along with rising operational costs, have made the increase in water price necessary to ensure the utility can cater to future demand, strengthen Singapore's water security, and continue delivering a high-quality and reliable supply of water.[2]

Table 13.1 Revised water price for domestic customers

	Phase 1: From 1 July 2017		Phase 2: From 1 July 2018	
Monthly water usage	0–40 m^3	>40 m^3	0–40 m^3	>40 m^3
Tariff	$1.19	$1.46	$1.21	$1.52
WCT	$0.42	$0.73	$0.61	$0.99
(% of tariff)	(35% of $1.19)	(50% of $1.46)	(50% of $1.21)	(65% of $1.52)
WBF	$0.78	$1.02	$0.92	$1.18
Total price	$2.39	$3.21	$2.74	$3.69

Table 13.2 Revised water price for non-domestic customers

	Phase 1: From 1 July 2017	Phase 2: From 1 July 2018
	Price ($/m^3)	Price ($/m^3)
Tariff	$1.19	$1.21
WCT (% of tariff)	$0.42 (35% of $1.19)	$0.61 (50% of $1.21)
WBF	$0.78	$0.92
Total price	$2.39	$2.74

Table 13.3 Revised water price for NEWater

	Phase 1: From 1 July 2017	Phase 2: From 1 July 2018
	Price ($/m^3)	Price ($/m^3)
Tariff	$1.28	$1.13
WCT (% of tariff)	$0.13 (10% of $1.19)	$0.13 (10% of $1.21)
WBF	$0.78	$0.92
Total price	$2.19	$2.33

Automated Meter Reading System Trial

PUB has been conducting a household trial on an Automated Meter Reading (AMR) system. The system allows PUB to transmit water consumption data wirelessly to users daily, allowing users to track, via an app, their water usage and take action to save water. The system also enables users to detect leaks early. The app enables users to see where high consumption is occurring for example in the shower and laundry and take steps to reduce water usage. The app has games that allow users to earn points and have chances to win prizes. Users also learn new water saving tips throughout the game itself. Households participating in the trial have so far achieved water savings of around 5% on average.[3]

Mandatory Water Efficiency Labelling Scheme

Started in 2009, the Mandatory Water Efficiency Labelling Scheme (Mandatory WELS) is a grading system with 0/1/2/3 tick rating denoting the water efficiency of a product. Currently, Mandatory WELS covers taps and mixers, dual-flush low capacity flushing cisterns, urinal flush valves and waterless urinals (1/2/3-tick rating) and washing machines (2/3/4-tick rating). It is mandatory for suppliers and retailers to obtain the relevant water efficiency labels for their products before advertising and displaying them for sale in Singapore. Overall, the scheme helps consumers make informed choices when making purchases. To read a WELS Label:

- Products with the most ticks are recommended
- The Label shows a product's water consumption, wash programme, type, brand, and model
- Each Label carries a registration number for validation.[4]

Water Efficiency Management Plan

Since January 2015, it is mandatory for all large water users who consume at least 60,000 cubic metres in the previous year to undergo a Water Efficiency Management Plan (WEMP) procedure, in which the customer must:

- Submit a notification to PUB by 31 March for all the different sites meeting this water consumption threshold
- Install private water meters at various water usage stages within their premises by 30 June to track and monitor water usage
- Submit their annual WEMP to PUB by 30 June for at least three consecutive years.

The WEMP allows customers to understand the breakdown of water usage in their premises and develop a Water Balance Chart, identify areas to further reduce consumption and raise efficiency, and establish an action plan that identifies measures in water savings, priorities, and implementation timelines.[5]

Water Efficiency Fund

Organisations that need to comply with WEMP can apply to PUB's Water Efficiency Fund to implement water-saving measures. The fund co-funds projects that yield at least a 10% reduction in water consumption within organisations.[6]

Water Efficient Building (Basic) Certification

Water Efficient Building (WEB) (Basic) Certification encourages businesses, industries, schools, and buildings to adopt water-efficient measures in their premises and procedures. By installing water efficient fittings and adopting WEB recommended flow rates/flush volumes (Table 13.4), premises can save around 5% of their monthly water consumption.[7]

Water Saving Habits

PUB's Internet resources provide a range of educational resources to encourage the public to conserve water and use it more efficiently, including the following:

Table 13.4 Water Efficient Building (Basic) water efficient flow rate/flush volume

Area of usage	Water efficient flow rate/flush volume
Basin tap and mixer	Two litres/minute (public/staff toilets) Four litres/minute (other areas)
Sink/kitchen tap and mixer	Six litres/minute
Shower tap and mixer and showerhead	Seven litres/minute
Bib tap and mixer	Six litres/minute
Urinal	0.5, 1, and 1.5 litres of water per flush for small, medium, and large size urinals respectively

Easy Steps to Remember Water-Saving Habits
To help the public translate knowledge to practice, PUB has rebranded its water-saving tips under the mnemonic W-A-T-E-R:

- WASH clothes on full load
- ALWAYS use half-flush when possible
- TURN off shower when soaping
- ENSURE tap is off when brushing teeth
- RINSE vegetables in container.

Simple Home Water Audit
PUB provides Internet resources, including a handbook, on how to conduct a simple home water audit that involves a five-step process:

1. Take meter reading and calculate water usage before the start of an audit
2. Check for leaks within your premises
3. Install thimbles if you are not using water efficient taps
4. Adopt water saving habits listed above
5. Re-take meter reading and calculate water savings after implementation.[8]

Reducing Leakage

PUB operates a few programmes to reduce leakage including the following:

Computerised Mains Replacement Programme
To reduce unaccounted-for-water, PUB uses a computerised system to capture information on mains including location, type, size, and age along with any details on previous leaks and repair work. The data is then used to plan the mains replacement programme where existing and potential problem areas are identified and prioritised for early replacement. One of the criteria used to determine if mains are due for replacement is the number of leaks occurring per kilometre per annum along sections of the water mains. Up to the end of the 1980s, a guideline of five leaks per kilometre per year was used. In the mid-1990s, PUB reduced this to three leaks per kilometre per year. Moving forward, PUB is reviewing the guideline to consider pipeline replacement based on two leaks per kilometre per year.[9]

Continuous Monitoring of the Network

PUB's Smart Water Grid involves 320 sensors deployed island-wide to monitor in real-time the pressure, flow, and quality of water in the network. This enables the utility to detect early leaks, pipe bursts, and water quality incidents, improve response time, and minimise the impact on customers. Moving forward, 120 sensors will be deployed on 100 kilometres of critical transmission mains for leak detection.[10]

DEVELOPING THE CIRCULAR WATER ECONOMY: REUSE AND RECYCLE

PUB has implemented a range of technology policies to increase the reuse and recycling of water resources in the development of the circular water economy.

NEWater

NEWater is the process of treating used water into ultra-clean, high-grade reclaimed water. PUB operates five NEWater plants that supply up to 40% of the country's needs. By 2060, NEWater is expected to meet up to 55% of Singapore's future demand. NEWater is used mainly for industrial and air-con cooling purposes at wafer fabrication plants, industrial estates, and commercial buildings. NEWater is delivered to industrial customers via a dedicated pipe network. To ensure it meets quality standards, NEWater undergoes a three-step treatment process involving microfiltration, reverse osmosis, and ultraviolet disinfection.[11]

Indirect Potable Reuse

During dry periods, NEWater is added to the reservoirs to blend with raw water. The raw water from the reservoir is then treated at the waterworks before being supplied to customers as tap water.

Deep Tunnel Sewerage System

The Deep Tunnel Sewerage System (DTSS) uses tunnel sewers to convey used water by gravity to centralised water reclamation plants (WRPs) located in the coastal areas. Treated used water is further purified into NEWater or discharged into the sea through outfalls. DTSS comprises a network of link sewers leading to two major tunnels (Phase 1 and 2) crisscrossing Singapore with three large WRPs at the northern (Kranji), eastern (Changi), and western (Tuas) ends of Singapore as well as outfall pipes.

Phase 1

Completed in 2008, Phase 1 of the DTSS comprises a 48-kilometre-long deep sewer tunnel running from Kranji to Changi, a centralised WRP at Changi, two five-kilometre-long deep-sea outfall pipes, and 60 kilometres of link sewers. The 'heart' of DTSS Phase 1 is the Changi WRP that can treat 900,000 cubic metres of used water per day.

Phase 2

In Phase 2, the DTSS will be extended by 2025 to serve the western part of Singapore. It will comprise a tunnel for conveying domestic used water and an industrial tunnel for non-domestic used water and associated link sewers. The advanced Tuas WRP will receive used water flows from the western part of Singapore via two tunnels: a tunnel to convey domestic used water and another to convey high-strength industrial used water, with both sources of used water to be treated separately. Domestic water will be treated at a 650,000 cubic metre/day module then purified into NEWater, while industrial water will be treated at a 150,000 cubic metre/day module to become industrial water and sent back to industries for reuse. Under DTSS Phase 2, a NEWater factory will be integrated with the Tuas WRP to facilitate water recycling, contributing to the goal of increasing NEWater supply from 40% to up to 55% of total water demand in the long term.

Phase 1 and Phase 2 Combined

The deep tunnels in Phase 2 will be connected to the existing deep tunnels in Phase 1 serving the eastern part of Singapore and the public sewer network to create one seamless integrated system. The whole of Singapore will be served by the DTSS when Phase 2 is completed by 2025. Used water will be conveyed from the DTSS via gravity to three centralised water reclamation plants for treatment, before it undergoes further treatment to produce NEWater, or discharged into the sea. When the entire DTSS is completed, existing intermediate pumping stations and conventional WRPs will be phased out, resulting in a 50% reduction in land taken up by used water infrastructure.[12]

Developing the Circular Water Economy: Recover

PUB has implemented a range of technology policies to recover resources in the development of the circular water economy.

Changi Water Reclamation Plant Recovering Energy

Changi WRP treats used water by removing solids and nutrients present in the used water. After treatment, the treated used water is safe to return to the environment or channelled to the NEWater factory to be further treated into highly purified NEWater. At Changi WRP, sludge is pumped to the anaerobic digesters for digestion with the biogas produced used as fuel for dryers to dry the sludge. The digested sludge is then dewatered in centrifuges to remove more water before it is further processed. The dewatered sludge from the dewatering centrifuges is sent to a series of rotary dryers. Biogas created in the digestion process runs the dryers, making them self-sufficient in energy.[13]

Tuas Nexus

Under DTSS Phase 2, PUB and the National Environment Agency (NEA) are building, by 2025, an integrated development comprising the Tuas WRP and the NEA's Integrated Waste Management Facility (IWMF), which combined will be known as 'Tuas Nexus'. The IWMF will be able to process the following waste streams:

- 5800 tonnes per day of incinerate waste
- 250 tonnes per day of household recyclables collected under the National Recycling Programme
- 400 tonnes per day of source-segregated food waste
- 800 tonnes per day of dewatered sludge from Tuas WRP.

The key benefits of IWMF are summarised in Table 13.5. The co-location of NEA's IWMF and PUB's Tuas WRP at the Tuas View Basin Site maximises both energy and resource recovery in solid waste and used water treatment processes, enabling NEA and PUB to benefit from a water-energy-waste nexus (Table 13.6).[14,15]

Table 13.5   Key benefits of the integrated waste management facility

Benefits	Description
Optimise land use	The IWMF will adopt an innovative plant layout and design to maximise the limited space available
Maximise energy and resource recovery	The IWMF will be designed for higher energy and resource recovery efficiencies. It will generate more electricity and recover more materials such as recyclables and metals for reuse
Minimise environmental impact	The IWMF will ensure clean air emissions as well as minimise solid residues for disposal at Semakau Landfill
Co-location synergies	The co-location of IWMF and Tuas WRP will enable both facilities to achieve higher plant performance and better cost efficiencies

Table 13.6   Key Tuas Nexus synergies

Synergies	Description
Material handling	• Food waste from IWMF to Tuas WRP for co-digestion with used water sludge
	• Dewatered sludge from Tuas WRP to IWMF for treatment and electricity production
	• Grit from Tuas WRP to IWMF for treatment
Energy	• Power supply from IWMF to Tuas WRP
	• Biogas from Tuas WRP to IWMF for higher overall thermal efficiency at IWMF
	• Steam from IWMF to Tuas WRP for sludge thermal hydrolysis and greasy waste treatment
Water	• Water from Tuas WRP to IWMF for process use
	• Used water from IWMF to Tuas WRP for treatment
Others	• Foul exhaust air from Tuas WRP to IWMF for combustion

Large-Scale Floating Solar Photovoltaic Systems

PUB is developing a 50-megawatt peak (MWp) large-scale floating solar photovoltaic (PV) system at Tengeh Reservoir that will be completed in 2021, providing around 7% of PUB's current energy needs. Separately, the Economic Development Board (EDB) is launching a Request for Information to explore the possibility of a 100 MWp floating solar PV system for private sector consumption, starting with studies at Kranji Reservoir.

Table 13.7 PUB case study summary

3R principle	Policy innovation	Description
Reduce	Components of the water price	The water price comprises a Water Tariff, covering production costs, a Water Conservation Tax, to encourage water conservation, and a Waterborne Fee for the cost of treating used water
	Water price revision	PUB has revised the water price for all customers upwards in two phases to ensure the utility can cater to future demand, strengthen water security, and continue delivering a high-quality and reliable supply of water
	Automated meter reading system trial	PUB has been conducting a household trial on an Automated Meter Reading system that allows users to track, via an app, their water usage and take action to save water
	Mandatory water efficiency Labelling scheme	The labelling scheme denotes the water efficiency of a product, helping consumers make informed choices when making purchases
	Water efficiency management plan	It is mandatory for all large water users to undergo a Water Efficiency Management Plan that involves installing private water meters, tracking and monitoring water usage, and submitting their annual plan
	Water efficiency fund	Organisations that need to comply with the Water Efficiency Management Plan can apply for funding to implement water saving measures
	Water efficient building (Basic) certification	Encourages businesses, industries, schools, and buildings to adopt water-efficient measures in their premises and procedures
	Water saving habits	PUB's Internet resources provide a range of educational resources to encourage the public to conserve water and use it more efficiently, including an easy-to-remember mnemonic 'W-A-T-E-R' that spells out how to conserve water at home as well as guidance on how to conduct a simple home water audit
	Reducing leakage	Computerised Mains Replacement Programme Computerised system captures information on mains along with any details on previous leaks and repair work to identify and prioritise mains replacements
		Continuous Monitoring of the Network PUB's Smart Water Grid enables early detection of leaks and pipe bursts
Reuse and recycle	NEWater	NEWater is the process of treating used water into ultra-clean, high-grade reclaimed water. During dry periods, NEWater is added to the reservoirs to blend with raw water
	Deep tunnel sewerage system	Tunnel sewers convey used water by gravity to centralised water reclamation plants located at the coastal areas. Treated used water is further purified into ultra-clean, high-grade reclaimed water or discharged into the sea through outfalls
Recover	Changi water reclamation Plant recovering energy	The biogas produced is used as fuel for dryers to dry the sludge. The digested sludge is then dewatered in centrifuges. The dewatered sludge from the dewatering centrifuges is sent to a series of rotary dryers. Biogas created in the digestion process runs the dryers, making them self-sufficient in energy
	Tuas Nexus	The facility maximises both energy and resource recovery in solid waste and used water treatment processes
	Large-scale floating solar photovoltaic systems	PUB is developing a large-scale floating solar photovoltaic system at Tengeh Reservoir and the Economic Development Board is exploring the possibility of a floating solar photovoltaic system for private sector consumption at Kranji Reservoir

Feasibility Studies
These two projects follow on from a feasibility study done at Tengeh Reservoir which involved PUB and EDB launching a 1MWp floating solar PV testbed to study the deployment of floating solar PV systems on Singapore's reservoirs. The systems were found to be 15% more efficient than conventional solar PV rooftop systems due to the cooler reservoir environment.[16]

CASE STUDY SUMMARY

PUB has implemented a range of 3R technology policies that facilitate the development of the circular water economy that not only mitigates greenhouse gas emissions but also enhances resilience to climate change (Table 13.7).

NOTES

1. PUB, "Water Supply," https://www.pub.gov.sg/watersupply.
2. "Water Price," https://www.pub.gov.sg/watersupply/waterprice.
3. *Automated Meter Reading (Amr) Trial* (2018).
4. "About Water Efficiency Labelling Scheme," https://www.pub.gov.sg/wels/about.
5. "Water Efficiency Management Plan," https://www.pub.gov.sg/savewater/atwork/managementplan.
6. "Efficiency Measures," https://www.pub.gov.sg/savewater/atwork/efficiencymeasures.
7. "Water Efficient Building (Basic) Certification," https://www.pub.gov.sg/savewater/atwork/certificationprogramme.
8. "Water Saving Habits," https://www.pub.gov.sg/savewater/athome/watersavinghabits.
9. R.C. Brears, *Urban Water Security* (Chichester, UK and Hoboken, NJ: Wiley, 2016).
10. Lai Kah Cheong, "Singapore's Experience of Water Loss Reduction," in *Water Loss Europe 2017* (Copenhagen, 2017).
11. PUB, "Newater," https://www.pub.gov.sg/watersupply/fournationaltaps/newater
12. "About Deep Tunnel Sewerage System," https://www.pub.gov.sg/dtss/about.
13. Ibid.
14. Ibid.

15. National Environment Agency, "Integrated Waste Management Facility," https://www.nea.gov.sg/our-services/waste-management/waste-management-infrastructure/integrated-waste-management-facility.
16. PUB, "Pub Pursues Large-Scale Floating Solar Deployment at Tengeh Reservoir While EDB Explores Potential for 100MWp System," https://www.pub.gov.sg/news/pressreleases/PUBpursueslargescalefloating solardeploymentatTengehReservoir.

References

Brears, R. C. *Urban Water Security.* Chichester, UK and Hoboken, NJ: Wiley, 2016.

Lai Kah Cheong. "Singapore's Experience of Water Loss Reduction." In *Water Loss Europe 2017.* Copenhagen, 2017.

National Environment Agency. "Integrated Waste Management Facility." https://www.nea.gov.sg/our-services/waste-management/waste-management-infrastructure/integrated-waste-management-facility.

PUB. "About Deep Tunnel Sewerage System." https://www.pub.gov.sg/dtss/about.

———. "About Water Efficiency Labelling Scheme." https://www.pub.gov.sg/wels/about.

———. *Automated Meter Reading (AMR) Trial.* 2018.

———. "Efficiency Measures." https://www.pub.gov.sg/savewater/atwork/efficiencymeasures.

———. "Newater." https://www.pub.gov.sg/watersupply/fournationaltaps/newater.

———. "Pub Pursues Large-Scale Floating Solar Deployment at Tengeh Reservoir While EDB Explores Potential for 100MWp System." https://www.pub.gov.sg/news/pressreleases/PUBpursueslargescalefloatingsolarde ploymentatTengehReservoir.

———. "Water Efficiency Management Plan." https://www.pub.gov.sg/savewater/atwork/managementplan.

———. "Water Efficient Building (Basic) Certification." https://www.pub.gov.sg/savewater/atwork/certificationprogramme.

———. "Water Price." https://www.pub.gov.sg/watersupply/waterprice.

———. "Water Saving Habits." https://www.pub.gov.sg/savewater/athome/watersavinghabits.

———. "Water Supply." https://www.pub.gov.sg/watersupply.

CHAPTER 14

Waternet Developing the Circular Water Economy

Abstract Waternet has implemented a range of technology policies to facilitate the development of the circular water economy that not only mitigates greenhouse gas emissions but also enhances resilience to climate change.

Keywords Water conservation · Water efficiency · Water reuse · Water recycling · Resource recovery · Renewable energy

INTRODUCTION

Waternet, responsible for the water management in and around Amsterdam, is the only water company in the Netherlands that covers the whole water cycle. With climate change impacting water resources across the Netherlands, Waternet is actively exploring a variety of solutions to adapt to climatic extremes and mitigate emissions. In addition, the City of Amsterdam, which is one of two owners of Waternet, is aiming to transition towards the circular economy involving all stakeholders. Considering this, the utility aims to reduce its energy consumption and explore innovative ways of working towards a green future. This chapter will discuss how Waternet has implemented a range of 3R technology policies (reduce, reuse and recycle, and recover) that facilitate

R. C. Brears, *Developing the Circular Water Economy*,
Palgrave Studies in Climate Resilient Societies,
https://Doi.org/10.1007/978-3-030-32575-6_14

the development of the circular water economy that not only mitigates greenhouse gas emissions but also enhances resilience to climate change.

DEVELOPING THE CIRCULAR WATER ECONOMY: REDUCE

Waternet has implemented a range of technology policies to enhance water conservation and promote water efficiency measures in the development of the circular water economy.

Tap Water Rates for Homes

The tap water rates for homes with water meters is composed of a volumetric rate as well as fixed charges to cover maintenance and administrative costs:

- €0.80 per m³ for your water use (1 m³ = 1000 litres)
- €0.343 per m³ tap water tax
- €73.81 fixed costs per year for the management and maintenance of the water mains.
- €1.50 for every bill sent by mail. Digitall bills are free of charge
- 9% Value Added Tax, levied over all costs listed above.[1]

Table 14.1 Education programmes for children and young people

Education programme	Description
Green linked	Anyone of any age can search for information on nature, sustainability, and education
Droppie water	Designed for elementary school students to learn about clean water, sanitation, flooding, and the work of water boards
Water wise	People can find everything about water management in the Netherlands. In TV clips, animations, games, etc., people can learn about how water is treated and about flood protection
Video: clean, safe, and adequate water	The animation shows what the duties of the Dutch water boards are and how they provide clean, safe, and adequate water
Water chats	A game about creating safe levees and providing sufficient clean water
Sieb the sheep	Sieb the sheep takes you into the world of rivers
Education ships	Children and young people can learn about water on the water itself
Guest lectures on water	Experts come in and tell the class about water in the Netherlands

School Education

For schools, Waternet has a special programme where children visit Waternet's plants and nature conservation areas and learn about water-related issues. In addition, Waternet has developed numerous education programmes that target school children in promoting awareness of water and sustainability, which are listed in Table 14.1.[2]

DEVELOPING THE CIRCULAR WATER ECONOMY: REUSE AND RECYCLE

Waternet has implemented a range of technology policies to increase the reuse and recycling of water resources in the development of the circular water economy.

Artificial Recharge in the Amsterdam Water Supply Dunes

Artificial recharge has been in operation in the dune area to the west of Amsterdam since 1957. The concept of the Amsterdam dune recharge system is based on using groundwater only for storage, removal of pathogens, and for attenuating peaks in water quality; integrating the recharge system within nature restoration goals; and preventing pollution of the water source. The objectives of the system are to stop overexploitation of the freshwater resources in the dune area, restore ecological conditions, and enhance the recharge capacity to over 65 million cubic metres per year to secure the water supply for the metropolitan region. The system of infiltration ponds and extraction means has been designed to have a minimum travel time of 60 days to remove pathogens. The capacity of the system has been designed to meet a two-month intake interruption. This is achieved by utilising the combined storage capacity of the shallow aquifers and wells tapping the deeper aquifers.

How It Works

The Amsterdam Water Supply Dunes cover an area of around 3500 hectares in the province Noord-Holland and produce around 60 to 65 million cubic metres of safe drinking water annually. Though supplemented by natural dune water, the main source of water is surface water from the Lek Canal, 55 kilometres away. The surface water is pre-treated close to the intake in Nieuwegein after which it is transported to the Amsterdam Water Supply Dunes. Once the water has percolated through the shallow groundwater system in the dunes at Vogelenzang, post-treatment takes place in the Leiduin production plant.[3]

Project SmartRoof 2.0

Waternet is conducting extensive research on blue-green roofs in Amsterdam, in particular, the cooling effect of evaporation and their use for storing and reusing rainwater for irrigation. The research roof is now open at the Marineterrein in Amsterdam under the name 'Project SmartRoof 2.0'. The roof provides multiple benefits including energy performance of the building, rainwater management, reduced urban heat island effect, and biodiversity. The project demonstrates the value of the combination of blue (collection and reuse of rainwater for irrigation) and green (varied planting) for a climate-proof and liveable city.[4]

DEVELOPING THE CIRCULAR WATER ECONOMY: RECOVER

Waternet has implemented a range of technology policies to recover resources in the development of the circular water economy.

New Sanitation at the Buiksloterham Pilot Project

Waternet is introducing New Sanitation at various locations in Amsterdam where dirty water from households is separated at the source to create two streams: blackwater and greywater. New Sanitation will be installed at the Buiksloterham pilot project as part of a wider circular economy project involving 24 participants. Vacuum toilets will be installed in Buiksloterham homes, replacing the ordinary toilets. A vacuum sewer will collect the blackwater while greywater will be collected with the normal sewer. The vacuum sewer will transport the blackwater to a digester in the district, enabling biogas to be utilised and phosphorous to be extracted. The existing sewer will collect greywater and because it is warm (on average more than 25 degrees Celsius) its heat can be recovered. This will involve linking a heat exchanger to a heat-cold storage unit in each house or apartment block.[5]

The Calcite Factory Under Construction

Waternet is constructing The Calcite Factory, a pilot project in the Western Harbour area of Amsterdam in partnership with Advanced Minerals, to process calcite pellets from softened drinking water into high-quality raw materials for customers in the food, feed, industry, and chemical sectors. The end product is intended to be an alternative to quarry chalk. Some of the calcite will be reprocessed into new seeding material used to soften water while the remaining calcite pellets will be processed to make high-quality niche products for ceramics, paints, soil improvers, and the carpet tile industry. Meanwhile, the company AquaMinerals will be responsible for sales. The pilot project intends for other drinking water companies to be involved when the factory becomes operational with the sector able to make its own seeding material, closing the supply chain as it will no longer need to purchase seeding material elsewhere. When the pilot plant is complete, each drinking water company will be able to supply (sand-free) calcite pellets and collect seeding material.

Harvesting Cellulose

Of all the raw materials in the sewage water, cellulose accounts for the greatest quantity. Waternet has begun a programme at several of its sewage treatment plants to measure the amount of cellulose discharged from the sewer. In addition, the utility is researching the most effective and cheapest way of harvesting cellulose. One project Waternet is participating in is the Waste2Aromatics consortium, which is attempting to convert cellulose fibres from a combination of organic waste, nappies, manure, and sieved waste into raw materials to produce aromatics. These can be used to make plastic.[6]

Biomass Is a Resource

Each year, Waternet harvests 6000 tonnes of dried aquatic plants, reed, and grass, which is normally disposed of as green waste or left to rot. The utility is now investigating the possibilities of reusing it with the aim to have the biocuttings reused and applied in products. By 2020, Waternet aims to reduce carbon dioxide emissions by seven kilotonnes with no additional costs.[7]

Biorefinery
With a biorefinery, biomass can be converted into components such as proteins and fibres. These can be used as raw materials for chemicals, biopolymers, and animal feed. For example, Waternet grew flax on the uncultivated land at the West sewage treatment plant, with it being used as a raw material to produce paint.

Biocomposite
Dried aquatic plants, reeds, and grasses can be used to make biocomposite. The glass fibre that is normally used to make composite can be replaced by fibre from plants. Biocomposites can be used to make products including furniture, sheet piling, and dashboards for cars.[8]

Heat from Shower Water

At the Uilenstede student accommodation in Amstelveen, Waternet installed 100 shower heat exchangers in one-person student flats. In 10 of the flats, continuous measurements were taken. The performance of the shower heat exchanges in the flats was compared to the test equipment in the Waternet test lab. The measurements will be taken until 2020 to observe long-term effects, including the build-up of soap residue in the shower heat exchanger. To date, some of the initial results include:

- The heat exchanger saves €50 worth of energy per person per year. With the cost of the equipment, including installation, around €600, the system has a payback period of 12 years in a single-person household and in less than five years with households with three people or more
- The heat exchangers do not require any maintenance for a significant amount of time
- Around half of the heat from the shower is fed back to the house with the average annual efficiency rate being more than 50%.[9]

Struvite from Urine

Waternet extracts phosphate from sewage water to make 1000 tonnes per year of struvite. There is a separate installation at Waternet's water treatment plant in Amsterdam Westpoort which produces 2500 kilograms of struvite per day. During special events, Waternet also collects urine from urinals with a special tank. The tank is sent to the sewage treatment plant in the Westelijk Havengebied where a separate installation converts the phosphate from the urine into struvite. With one litre of pure urine providing two grams of struvite, the Televizier-Ring gala in 2016 collected 550 litres of urine to produce 1.1 kilograms of struvite.[10]

Reuse of Paper from Wastewater

At the sewage treatment plant in Hilversum, Waternet tested a new sieve and found a way to filter paper from wastewater. The paper fibres can be used to make asphalt. Waternet is also investigating whether the fibres are suitable to produce, for example, clothing. Whether there is a large-scale roll-out of paper fibres is dependent on:

- *The cost*: The costs of extracting cellulose needs to be reduced either by cooperation with other utilities or applying new sieving techniques
- *How much it yields*: There needs to be an assessment on the financial yield of cellulose from sieves, what investment is required in the screening plant, and whether an installation is needed to clean the fibres
- *The effect*: Is the quality of purified wastewater improved? What is the source of energy? And what is the effect on the purification process?[11]

Waste-to-Energy

Waternet's main wastewater treatment plant 'Amsterdam West' is located beside a waste-to-energy plant operated by AEB Waste to Energy Company. The proximity enables an exchange of energy flows between the two plants with large environmental benefits: Amsterdam West produces 25,000 cubic metres per day of biogas and 100,000 tonnes of sewage sludge per year for burning at the waste-to-energy plant. The energy produced in the waste-to-energy plant is then used to power the Amsterdam West treatment plant. In total, the integration of the two plants produces 20,000-megawatt hours per year of electricity and 50,000 gigajoules per year of heat, saving 1.8 million cubic metres per year of natural gas, resulting in avoided greenhouse gas emissions of 3200 tonnes per year.[12]

100,000 Solar Panels

Waternet is aiming to have 100,000 solar panels installed by 2020. To date, the utility is assessing the best locations for the panels with research conducted on 34 locations. With 100,000 panels, the utility would reduce carbon dioxide emissions by 10,000 tonnes, which is 20% of the goal of being carbon-neutral by 2020. Around 88% of the panels will be installed on the ground with the remaining on the roofs. In total, the 100,000 panels would produce 26,000 kWp and yield 23,000,000 kWh, equivalent to the electricity use of 6500 households per annum.

CASE STUDY SUMMARY

Waternet has implemented a range of 3R technology policies that facilitate the development of the circular water economy that not only mitigates greenhouse gas emissions but also enhances resilience to climate change (Table 14.2).

Table 14.2   Waternet case study summary

3R principle	Policy innovation	Description
Reduce	Tap water rates for homes with meters	Waternet's tap water rates for homes with meters is comprised of a volumetric rate and fixed charges
	School education	Waternet has a special programme where children visit Waternet's plants and nature conservation areas and learn about water-related issues. In addition, Waternet has developed numerous education programmes that raise awareness of water and sustainability
Reuse and recycle	Artificial recharge in the Amsterdam water supply dunes	The objectives of the system are to stop overexploitation of the freshwater resources in the dune area, restore ecological conditions, and enhance the recharge capacity to secure the water supply for the metropolitan region
	Project SmartRoof2.0	Waternet is conducting extensive research on blue-green roofs in Amsterdam, including their use for storing and reusing rainwater for irrigation
Recover	New sanitation	The Buiksloterham pilot project will have vacuum toilets collecting the blackwater, enabling biogas to be utilised and phosphorous extracted, while greywater will be collected in the normal sewer and its heat recovered
	The Calcite Factory under construction	Waternet is constructing, in partnership with Advanced Minerals, The Calcite Factory to process calcite pellets from softened drinking water into high-quality raw materials for customers in the food, feed, industry, and chemical sectors

3R principle	Policy innovation	Description	
Harvesting cellulose		Waternet is participating in a consortium that is attempting to convert cellulose fibres from waste into raw materials to produce aromatics. These can be used to make plastic	
Biomass is a resource	*Biorefinery*	Waternet grew flax on uncultivated land at its sewage treatment plant for use as a raw material in producing paint	
	Biocomposite	Waternet is exploring the use of biocomposites in making a variety of products including furniture, sheet piling, and dashboards for cars	
Heat from shower water		Waternet has installed 100 shower heat exchangers in one-person student flats to test their performance	
Struvite from urine		Waternet extracts phosphate from sewage water to make struvite. There is a separate installation at Waternet's water treatment plant in Amsterdam Westpoort which produces struvite. During special events, Waternet also collects urine from urinals to produce struvite	
Reuse of paper from wastewater		Waternet tested a new sieve and found a way to filter paper from wastewater. The paper fibres can be used to make asphalt. Waternet is also investigating whether the fibres are suitable to produce clothing	
Waste-to-energy		Waternet's main wastewater treatment plant is located beside a waste-to-energy plant. The proximity enables an exchange of energy flows between the two plants with large environmental benefits	
100,000 solar panels		Waternet is aiming to have 100,000 solar panels installed by 2020. To date, the utility is assessing the best locations for the panels with research conducted on 34 locations	

NOTES

1. Waternet, "Tap Water Rates for Homes with a Water Meter," https://www.waternet.nl/en/service-and-contact/tap-water/costs/rates-with-a-water-meter/.
2. R.C. Brears, *Urban Water Security* (Chichester, UK and Hoboken, NJ: Wiley, 2016).
3. Partners for Water Programme, "The Amsterdam Dune Water Machine," (2014), http://metameta.nl/wp-content/uploads/2016/09/1_Amsterdam_Water_supply_Dunes_EN.pdf.
4. Waternet, "Onderzoek & Innovatie," (2017), https://www.waternet.nl/globalassets/innovatie-platform-content--media/innovatie-homepage/jaarverslagen-onderzoek-en-innovatie/oi-jaarverslag-2017-web.pdf.
5. Ibid.
6. Ibid.
7. Ibid.

8. Ibid.
9. "The Second Life of Waste Water," https://www.waternet.nl/blog/tweede-leven-afvalwater/.
10. "Struvite from Urine," https://www.agv.nl/onze-taken/klimaatproblemen-aanpakken/struviet-uit-urine/.
11. "We Can Reuse Paper from Waste Water," https://www.agv.nl/blog/hoe-we-papier-terugwinnen-uit-afvalwater-om-opnieuw-te-gebruiken/.
12. J. P. Van der Hoek, A. Struker, and J. E. M. Danschutter, "Amsterdam as a Sustainable European Metropolis: Integration of Water, Energy and Material Flows," in *International Water Week* (Amsterdam, The Netherlands, 2013).

REFERENCES

Brears, R. C. *Urban Water Security.* Chichester, UK and Hoboken, NJ: Wiley, 2016.
Partners for Water Programme. "The Amsterdam Dune Water Machine." (2014). http://metameta.nl/wp-content/uploads/2016/09/1_Amsterdam_Water_supply_Dunes_EN.pdf.
Van der Hoek, J. P., Struker, A. and Danschutter, J. E. M. "Amsterdam as a Sustainable European Metropolis: Integration of Water, Energy and Material Flows." In *International Water Week.* Amsterdam, The Netherlands, 2013.
Waternet. "Onderzoek & Innovatie" (2017). https://www.waternet.nl/globalassets/innovatie-platform-content--media/innovatie-homepage/jaarverslagen-onderzoek-en-innovatie/oi-jaarverslag-2017-web.pdf.
———. "The Second Life of Waste Water." https://www.waternet.nl/blog/tweede-leven-afvalwater/.
———. "Struvite from Urine." https://www.agv.nl/onze-taken/klimaatproblemen-aanpakken/struviet-uit-urine/.
———. "Tap Water Rates for Homes with a Water Meter." https://www.waternet.nl/en/service-and-contact/tap-water/costs/rates-with-a-water-meter/.
———. "We Can Reuse Paper from Waste Water." https://www.agv.nl/blog/hoe-we-papier-terugwinnen-uit-afvalwater-om-opnieuw-te-gebruiken/.

CHAPTER 15

Best Practices

Abstract From the case studies a series of best practices have been identified that can be implemented by other locations around the world
attempting to develop the circular water economy.

Keywords Water conservation · Water efficiency · Water reuse ·
Water recycling · Resource recovery · Renewable energy

INTRODUCTION

From the case studies of water utilities implementing a range of 3R
technology policies (reduce, reuse and recycle, and recover), a series of
best practices have been identified in enhancing water conservation and
promoting water efficiency, reusing and recycling water resources, and
recovering resources in the development of the circular water economy.
These best practices can be implemented by other locations around the
world attempting to develop the circular water economy.

© The Author(s) 2020   195
R. C. Brears, *Developing the Circular Water Economy*,
Palgrave Studies in Climate Resilient Societies,
https://doi.org/10.1007/978-3-030-32575-6_15

Developing the Circular Water Economy: Reduce

From the case studies, a series of best practices have been identified of water utilities implementing a range of technology policies to promote water conservation and enhance water efficiency measures in the development of the circular water economy.

Water Pricing

Water utilities use a variety of water pricing structures to encourage water conservation including the following:

- New York City's Department of Environmental Protection (DEP) charges a flat-rate for water consumption in residential units that are equipped with Automatic Meter Reading (AMR) devices and have completed water efficiency measures
- If San Francisco Public Utilities Commission (SFPUC) adopts a resolution declaring a stage of water delivery reduction, a drought surcharge is applied to water rates
- Singapore's Public Utilities Board (PUB) has a water price that comprises a Water Tariff, covering production costs, a Water Conservation Tax, to encourage water conservation, and a Waterborne Fee for the cost of treating used water. PUB has revised the water price for all customers upwards in two phases to ensure the utility can cater to future demand, strengthen water security, and continue delivering a high-quality and reliable supply of water.

Regulations

Water management generally comes in the form of temporary and permanent regulations:

- Austin Water requires:
  - Large commercial, multi-family, and City of Austin properties complete an irrigation system inspection every two years
  - Commercial, multi-family, and city municipal facilities with vehicle washes using potable water submit an annual efficiency evaluation
  - All properties with cooling towers register them with Austin Water so potential water-saving upgrades can be identified and owners kept informed of available rebates

- PUB's mandatory water efficiency labelling scheme helps consumers make informed choices when making purchases
- It is mandatory for all of PUB's large water users to undergo a Water Efficiency Management Plan that involves installing private water meters, tracking and monitoring water usage, and submitting their annual plan.

Subsidies, Grants, and Rebates

Subsidies, grants, and rebates are used to modify water users' behaviour in a predictable, cost-effective way, that is, reduce wastage and lower water consumption. A variety of best practices are as follows:

- Austin Water offers residents and commercial/multi-family/school customers a variety of rebates for water-efficient devices
- DEP has a voucher-based programme that provides a discount to customers on the cost of new toilet fixtures
- DEP provides funding for water demand reduction projects in City-owned facilities
- SFPUC provides grants to non-residential customers who can significantly reduce their use of potable water through upgrades or replacement of existing onsite indoor water-using equipment.

Free Advice and Evaluations

Free advice and evaluations raise awareness on how to conserve water in a variety of settings:

- Austin Water has contracted Dropcountr Inc. to offer customers free, digital home water use reports
- Austin Water provides customers with free irrigation system evaluations to become more efficient and identify any need for repairs and upgrades
- SFPUC offers free evaluations to help customers identify leaks and replace inefficient fixtures. Tips are also provided to help customers use indoor and outdoor water more efficiently
- SFPUC can help retail water users identify water saving opportunities in their irrigation systems.

Free Water-Saving Kits

Free water-saving kits can be distributed along with tips to help conserve water. A variety of best practices are as follows:

- Anglian Water provides water-saving home visits in which a utility representative will provide water saving advice and fit water-saving products where possible
- Anglian Water offers free garden water saving kits to customers to conserve water while still having a beautiful garden
- Austin Water offers free tools to help residential customers save water both indoors and outdoors.

Education

Education of the public is crucial in generating an understanding of water scarcity and creating acceptance of the need to implement water conservation programmes.

Water utilities can also promote water conservation in schools to increase young people's knowledge of the water cycle and encourage the sustainable use of scarce water resources. A variety of best practices are as follows:

- Anglian Water's purpose-built education centres at its water recycling centres provide an informative educational experience for students on water conservation
- South Australia Water Corporation (SA Water) produces a range of educational resources for use in classrooms including books, toys, and videos
- SA Water holds free events and distributes learning resources that support the Australian Curriculum in the areas of geography, science, and sustainability
- PUB's Internet resources provide a range of educational resources to encourage the public to conserve water and use it more efficiently, including an easy-to-remember mnemonic 'W-A-T-E-R' that spells out how to conserve water at home. In addition, PUB provides guidance on how to conduct a simple home water audit
- Waternet has a special programme where children visit Waternet's plants and nature conservation areas and learn about water-related

issues. In addition, Waternet has developed numerous education programmes that raise awareness of water and sustainability.

Educational Smartphone Apps

Educational smartphone apps widen the reach of education programmes. For instance, SA Water has developed smartphone apps to encourage water conservation including a real-time management game that puts the player in control of the water and the population's happiness.

Smart Meters and Online Portals

Smart meters and online portals enable water users to directly relate their water bill with water consumption. A variety of best practices include the following:

- Anglian Water is trialling smart meters in several communities, with the smart meters transmitting hourly readings from the property with customers able to access it via their 'My Use' portal
- DEP customers can log into their My DEP Account to view and manage their consumption
- DEP customers can use any Amazon Alexa-enabled device to track their water usage, account balance, payments, and bills
- DEP proactively alerts customers to potential water leaks on their property
- SA Water has implemented a new online portal that lets customers view their water use, with comparison data available
- SA Water has begun a 12-month smart meter pilot programme with flow and pressure sensors installed along with smart meters for residential and business customers
- PUB has been conducting a household trial on an AMR system that allows users to track, via an app, their water usage and take action to save water.

Services for Businesses to Reduce Water Usage

A variety of best practices have been identified of utilities encouraging businesses to reduce water for both domestic and non-domestic purposes:

- Anglian Water Business:
  - Works with customers to develop an effective water management strategy to identify and address any water risks that could impact operational efficiency
  - Offers customers an AMR Service to reduce water usage and improve efficiency
  - Provides a service to help business customers reduce onsite domestic water use
  - Helps customers identify potential optimisation opportunities and water savings across all site-based production and operational processes
  - Provides a water audit service for agricultural customers to improve their water efficiency. The audit can be used as a benchmark to assess performance against other agricultural businesses. Anglian Water Business can also help agricultural customers understand the risks of water scarcity and protect their business
- SA Water's Business Relations team can help business customers find the best ways to improve water efficiency and find leaks.

Water Conservation Challenges

Water utilities have initiated a range of competitive water conservation challenges to a variety of large water-using sectors, including the following:

- Austin Water encourages businesses to learn ways of reducing water use by 10%, which can lower energy and wastewater costs. The utility offers free recognition for lodging facilities that use water-efficient measures and practices
- DEP has issued the New York City Water Challenge to Universities to implement water conservation strategies and foster a water conservation ethic among students
- PUB's Water Efficient Building (Basic) Certification programme encourages businesses, industries, schools, and buildings to adopt water-efficient measures in their premises and procedures.

Reducing Leakage

A variety of leak reduction initiatives have been implemented to reduce non-revenue water. A couple of best practice examples are:

- Anglian Water's Optimised Water Networks Strategy aims to prevent bursts and leaks through better management of the pressure in the network
- PUB's computerised system captures information on mains along with any details on previous leaks and repair work to identify and prioritise mains replacements. Also, PUB's Smart Water Grid enables early detection of leaks and pipe bursts.

DEVELOPING THE CIRCULAR WATER ECONOMY: REUSE AND RECYCLE

From the case studies, a series of best practices have been identified of water utilities implementing a range of technology policies to increase the reuse and recycling of water resources in the development of the circular water economy.

Financial Incentives

Water utilities have implemented a variety of financial incentives to encourage water reuse and water recycling, of which a variety of best practices include:

- Anglian Water encourages developers to install Green Water systems in new homes including rainwater harvesting systems, stormwater harvesting systems, and water recycling systems with homebuilders receiving a waiver on zonal charges for building sustainable homes that incorporate Green Water systems
- DEP provides grants to customers who install rainwater, blackwater, or greywater systems at both the building-scale and district-scale. Customers who install water reuse systems that reduce the building's water consumption by at least 25% also receive a 25% fee discount
- SFPUC:
  - Provides grants to encourage property owners to design, build, and maintain performance-based green stormwater infrastructure including rainwater harvesting systems for non-potable reuse
  - Offers residents a discount off the purchase of a Laundry-to-Landscape greywater kit that includes the basic components to divert clothes washer water to gardens

- Offers a rebate to help cover the costs of greywater projects that require a permit from the Department of Building Inspection
- Provides a grant for retail water users to collect, treat, and use alternate water sources for non-potable uses.

Regulations

A couple of water utilities have implemented regulations to increase the uptake of water reuse and water recycling systems:

- Austin Water's price of reclaimed water is around one-third of the price of drinking water. City Code requires all new commercial locations within 250 feet of a reclaimed water main to connect for non-potable water uses
- SFPUC requires:
  - That all new large development projects, that have not received a site permit, install and operate an onsite non-potable water system to treat and reuse greywater, rainwater, and foundation drainage for toilet and urinal flushing and irrigation
  - Property owners to install recycled water systems in new construction, modification, or remodel projects.

Informational Resources

To facilitate the uptake of water reuse and water recycling systems, Austin Water provides Internet resources for customers considering onsite water reuse systems. Austin Water also provides help to property owners in the permitting process.

Rainwater Harvesting

To encourage the uptake of rainwater harvesting systems, a couple of best practices include:

- DEP giving away rain barrels, which can be used to water lawns and gardens or for other non-potable uses
- Waternet conducting extensive research on blue-green roofs in Amsterdam, including their use for storing and reusing rainwater for irrigation.

Recycled Water Infrastructure

A couple of water utilities have implemented recycled water infrastructure in the development of the circular water economy:

- SA Water's recycled water is supplied on a separate system using purple lines. It has been treated to a standard that is safe for a range of household purposes. SA Water is also increasing its production of recycled irrigation water by 60%, delivering climate and season independent water to the farm gate
- PUB's NEWater is created from the process of treating used water into ultra-clean, high-grade reclaimed water
- PUB's tunnel sewers convey used water by gravity to centralised water reclamation plants located at the coastal areas. Treated used water is further purified into NEWater or discharged into the sea through outfalls.

Indirect Potable Reuse

During dry periods, PUB's NEWater is added to the reservoirs to blend with raw water.

Groundwater Projects

From the case studies a couple of best practices of using groundwater to supplement potable water supplies have been identified:

- SFPUC treats and blends groundwater with the utility's regional drinking water supplies before it is delivered to customers
- Waternet uses artificial recharge to stop overexploitation of the freshwater resources in the dune area, restore ecological conditions, and enhance the recharge capacity to secure the water supply for the metropolitan region.

Environmental Reuse

From the case studies a couple of environmental reuse best practices have been identified:

- Anglian Water has created a natural treatment plant for over a million litres of water a day to help improve the quality of water that is returned to a river. Used but treated water passes through a wetland to be further filtered and cleaned before it is returned to the river. In addition, the wetland is a significant biodiversity asset for the region
- SFPUC's Green Infrastructure Grant Program encourages property owners to design, build, and maintain performance-based green stormwater infrastructure including rainwater harvesting systems that provide multiple benefits including non-potable reuse and groundwater recharge.

DEVELOPING THE CIRCULAR WATER ECONOMY: RECOVER

From the case studies, a series of best practices have been identified of water utilities implementing a range of technology policies to recover resources in the development of the circular water economy.

Biogas and Combined Heat and Power

A number of water utilities have implemented biogas and combined heat and power (CHP) systems, including the following:

- Anglian Water operates CHP engines. It is also trialling the use of E-STOR technology that uses second life Renault electric vehicle batteries to provide a smart, affordable, and flexible approach to grid load management
- All of Austin's sewage solids are pumped to the Hornsby Bend Wastewater Treatment Plant. Methane gas produced in the treatment process is recycled to generate electricity and heat. The biogas generator can generate more power than is needed to run the treatment plant. The excess electricity produced is fed back into the electric grid
- Part of the energy that powers SFPUC's Oceanside and Southeast wastewater treatment plants is generated from biogas. The biogas-fuelled cogeneration technology provides both heat and electricity for plant operations
- Work is underway to purify the biogas generated at DEP's Newtown Creek Wastewater Treatment Plant into pipeline-quality renewable natural gas for direct in-home use

- At PUB's Changi Water Reclamation Plant, the biogas produced is used as fuel for dryers to dry the sludge. The digested sludge is then dewatered in centrifuges. The dewatered sludge from the dewatering centrifuges is sent to a series of rotary dryers. Biogas created in the digestion process runs the dryers, making them self-sufficient in energy.

Biosolids

Several water utilities have implemented biosolid recovery projects, including the following:

- Austin Water's biosolids are anaerobically digested and composted into an Environmental Protection Agency-certified soil conditioner which is donated to landscape public spaces and sold to commercial vendors
- Biosolids are available free of charge from selected SA Water wastewater treatment plants
- SFPUC's Southeast Treatment Plant aims to produce higher quality biosolids and maximise biogas utilisation and energy recovery to produce heat, steam, and energy
- SFPUC has entered into a public-private partnership to sell biofertiliser to farms and ranches in California.

Recovery of Resources

A number of resource recovery initiatives have been undertaken to further the development of the circular water economy, including:

- SFPUC runs a programme to collect fats, oils, and grease to prevent it from being disposed of in the sewage system. After removal of impurities and primary polishing, SFPUC sells the grease by-product for conversion into biodiesel. The biofuel can then be sold to city transit fleets to offset their carbon emissions from fossil fuels
- Waternet:
  - Is involved in the construction of The Calcite Factory to process calcite pellets from softened drinking water into high-quality raw materials for customers in the food, feed, industry, and chemical sectors

- Is participating in a consortium that is attempting to convert cellulose fibres from waste into raw materials to produce aromatics. These can be used to make plastic
- Extracts phosphate from sewage water to make struvite. There is a separate installation at Waternet's water treatment plant in Amsterdam Westpoort which produces struvite. During special events, Waternet also collects urine from urinals to produce struvite
- Tested a new sieve and found a way to filter paper from wastewater. The paper fibres can be used to make asphalt. Waternet is also investigating whether the fibres are suitable to produce clothing
- Has implemented projects to recover resources from biomass, including growing flax on uncultivated land at its sewage treatment plant for use as a raw material in producing paint and exploring the use of biocomposites that can make a variety of products including furniture, sheet piling, and dashboards for cars.

Thermal Energy Recovery

A couple of water utilities have implemented thermal energy recovery systems:

- PUB's Tuas Nexus will involve biogas from PUB's water reclamation plant used to power the integrated waste management facility nearby, meanwhile steam from the waste management facility will be used at the water reclamation plant for waste treatment
- Waternet has installed 100 shower heat exchangers in one-person student flats to test their performance
- Waternet's Buiksloterham pilot project will have vacuum toilets collecting the blackwater, enabling biogas to be utilised and phosphorous extracted, while greywater will be collected in the normal sewer and its heat recovered
- Waternet's main wastewater treatment plant is located beside a waste-to-energy plant. The proximity enables an exchange of energy flows between the two plants with large environmental benefits.

Solar Energy

The following water utilities have installed or are exploring the installation of solar panels on facility-owned infrastructure:

- Anglian Water will install solar panels on land at one of its sites to generate renewable energy
- New York City's Port Richmond Wastewater Treatment Plant has one of the city's largest solar arrays. It was built through a public-private partnership with no upfront capital cost to the City, with Con Edison Solutions owning and maintaining the solar array and the City purchasing the electricity
- At an SFPUC wet weather facility, two separate building roofs are used to provide supplemental solar power. Meanwhile, a solar photovoltaic (PV) system has been installed on the roofs of two wastewater treatment buildings at a plant
- Waternet is aiming to have 100,000 solar panels installed by 2020. To date, the utility is assessing the best locations for the panels with research conducted on 34 locations.

Reservoir Solar Arrays

Several water utilities are installing or have installed floating solar arrays on their reservoirs:

- SA Water will trial floating solar panel arrays on its Happy Valley Reservoir with the installation of a 100-kilowatt pilot system
- SFPUC has installed the city's largest solar installation on one of its reservoirs, with excess renewable energy used to help power public buses, the San Francisco International Airport, health clinics, and other city services
- PUB is developing a large-scale floating solar PV system at Tengeh Reservoir and the Economic Development Board is exploring the possibility of a floating solar PV system for private sector consumption at Kranji Reservoir.

Energy Storage

A couple of water utilities have installed energy storage systems:

- Anglian Water is working with commercial partners on an integrated energy storage project that will see an energy storage machine installed alongside a solar PV system at a water treatment works

- SA Water is trialling a solar PV and battery storage system at its Crystal Brook workshop. This is in addition to SA Water trialling new silicon thermal energy storage technology at a wastewater treatment plant.

Wind Power

Anglian Water currently has three wind turbines installed and will continue to seek appropriate sites for future wind turbines.

Hydropower

San Francisco's drinking water is harnessed to generate hydroelectric power that meets about 17% of the City's electricity needs.

CHAPTER 16

Conclusions

Abstract The development of the circular water economy, which mitigates greenhouse gas emissions and enhances resilience to climate change, can be guided by innovative technology policies that encourage the reducing of water consumption, reusing and recycling of water, and recovery of materials from wastewater.

Keywords Water conservation · Water efficiency · Water reuse · Water recycling · Resource recovery · Renewable energy

In the linear economy, the water sector typically employs the Take-Use-Discharge strategy where water is 'withdrawn' from streams, rivers, lakes, reservoirs, oceans, and groundwater reservoirs as well as harvested directly as rainwater. Water is then 'used' by municipalities, industries, agriculture, the environment, etc. within the water cycle, including for consumptive and non-consumptive uses. Non-consumptive used water is 'returned' to the river basin directly or via a municipal treatment facility. Depending on the location within the basin this returned water could then be used downstream or lost to the basin.

In response to a variety of climatic and non-climatic challenges, water utilities are beginning to promote the development of the circular water economy, based on the 3R's of reducing water consumption, reusing and recycling water, and recovering materials from wastewater to not only mitigate greenhouse gas emissions but also enhance resilience to climate change.

In the circular water economy, the concept of reduce is achieved through demand management approaches that encourage using less water and making sure water is used efficiently. Demand management promotes water conservation and efficiency during both normal and abnormal conditions, through changes in practices, culture, and people's attitudes towards water resources. Demand management seeks to reduce the loss and misuse of water, optimise the use of water, and facilitate major financial and infrastructural savings by minimising the need to meet increasing demand with new water supplies. There are a variety of demand management tools available to water utilities to promote water conservation and enhance water efficiency including pricing of water, installation of water meters, economic instruments such as subsidies or rebates, temporary and permanent regulations to restrict certain types and levels of water use, water efficiency labelling of household appliances, and education and awareness campaigns. Water utilities can also reduce leakage in the water distribution system as well as promote the use of natural infrastructure to enhance water conservation and lower operating costs of treating wastewater.

Water reuse is defined as the reuse of water within a single process or the use of harvested water for another purpose without treatment, while recycle is defined as the use of harvested water for another purpose, after treatment. Water reuse and water recycling are often utilised for industrial processes, agricultural production, urban agriculture, indirect potable reuse, direct potable reuse, and supplementing aquifer levels and stream and river flows. In urban areas, onsite non-potable water systems are utilised to meet non-potable needs, with systems using greywater, blackwater, rainwater, and stormwater for cooling buildings, irrigating landscapes, and flushing toilets and urinals.

In the circular water economy, wastewater treatment plants are resource recovery facilities that in addition to producing clean water generate renewable energy and recover nutrients. The built infrastructure of water and wastewater treatment plants also provide opportunities to generate renewable energy. Energy derived from wastewater treatment is a renewable energy resource and can include electrical energy, heat or biofuels from utilisation of digester gas, electrical energy and heat from thermal conversion of biomass, electrical energy from biosolids products used by other entities, and heating or cooling energy using plant influent or effluent as a heat source or sink. Water utilities can also implement traditional renewable energy activities including the use of solar photovoltaic (PV) systems installed on facility buildings, floating solar PV plants installed on reservoirs, installation of on-site wind turbines, and energy recovery from hydropower. Finally, in the circular water economy, wastewater is a resource with water utilities able to recover nitrogen and phosphorous for reuse, create biosolids for use as a fertiliser, produce biodegradable plastic from wastewater, manufacture bricks from incinerated sewage sludge, and mine wastewater for metals.

From the case studies of Anglian Water, Austin Water, New York City Department of Environmental Protection, South Australia Water Corporation, San Francisco Public Utilities Commission, Singapore's Public Utilities Board, and Waternet, a series of best practices have been identified in enhancing water conservation and promoting water efficiency, reusing and recycling water resources, and recovering resources in the development of the circular water economy.

Water utilities are implementing a range of technology policies to promote water conservation and enhance water efficiency measures in the development of the circular water economy. Water pricing is used in a variety of ways including charging a flat-rate for residential customers equipped with automatic meter readers, adding a drought surcharge during times of scarcity, and raising water prices to ensure future water demand is met. Regulations are used to reduce water use and enhance efficiency including mandatory water labelling schemes on household products and devices and the requirement that all large commercial water users implement water efficiency plans and install private meters. Subsidies, grants, and rebates are used to reduce wastage and enhance efficiency, examples of which include rebates for water-efficient devices, vouchers for discounts on new water-efficient toilets, funding for city-owned buildings to implement water reduction initiatives, and grants for commercial customers to upgrade or replace existing onsite indoor water-using equipment. Free advice and evaluations are frequently offered to both domestic and commercial customers including evaluations of irrigation systems and water audits to identify leaks. Water utilities are distributing free water-saving kits to help homeowners save water both indoors and outdoors as well as offering digital tools that provide free reports on home water-use. Education programmes are used to generate an understanding of water scarcity and create acceptance of water conservation programmes, examples of which include purpose-built

educational centres at wastewater treatment plants to provide an informative experience for students, distribution of educational resources for
use in classrooms, and Internet resources that provide guidance on how
to do a simple water audit at home. Smartphone apps also widen the
reach of education programmes, with games devised to encourage water
conservation and resource efficiency. In the home, smart meters and
online portals provide alerts for leaks as well as enable the comparison of
water usage with others. To reduce leakage, water utilities are developing
computerised systems to identify and prioritise pipe replacements as well
as using smart water grids to detect leaks and pipe bursts.

Water utilities are implementing a range of technology policies to
increase the reuse and recycling of water resources in the development
of the circular water economy. Water utilities have deployed a variety of
financial incentives including offering homebuilders waivers on administration charges when installing rainwater harvesting systems, stormwater
harvesting systems, and water recycling systems; grants to customers who install rainwater, blackwater, or greywater systems at both the
building-scale and district-scale; grants to encourage property owners to
design, build, and maintain rainwater harvesting systems; discounts off
the purchase of greywater systems; and rebates to help cover the costs
of installing water reuse systems for non-potable uses. Water utilities
have implemented regulations to increase the uptake of water reuse and
water recycling systems including requiring all new commercial locations within areas serviced by reclaimed water to connect for non-potable water uses; requiring all new developments to install and operate
onsite non-potable water systems to treat and reuse greywater, rainwater, and foundation drainage for toilet and urinal flushing and irrigation; and requiring property owners to install recycled water systems in
new developments or upgrades. To facilitate the uptake of water reuse
and water recycling systems, Internet resources have been developed to
inform customers of the various types of systems available. To encourage
the uptake of rainwater harvesting systems, one utility is giving away rain
barrels while another is conducting research on the possibility of using
rooftops to store and reuse rainwater for irrigation. Water utilities have
implemented recycled water infrastructure including supplying domestic water users with recycled water on separate systems for non-potable
uses, building new water treatment facilities to deliver climate and season independent recycled irrigation water to the farm gate, and creating
infrastructure to turn used water into ultra-clean, high-grade reclaimed

water for industrial use. Water utilities have initiated groundwater projects to supplement water supplies, such as treating and blending groundwater with regional drinking water supplies before it is delivered to customers. Utilities are also encouraging environmental reuse of water with one utility having created a natural treatment plant to help improve the quality of treated wastewater being returned to the environment, with significant biodiversity benefits for the region, while another utility's green infrastructure grant programme encourages property owners to manage stormwater for a variety of benefits including non-potable reuse and groundwater recharge.

Water utilities are implementing a range of technology policies to recover resources in the development of the circular water economy. A few water utilities have implemented biogas and combined heat and power systems, including a utility that is storing its generated energy in electric vehicle batteries. Water utilities have implemented biosolid recovery projects to generate certified soil conditioner and biofertiliser for farms and ranches. A number of additional resource recovery initiatives have been undertaken to further develop the circular water economy, including processing calcite pellets into high-quality raw materials for customers in food, feed, industry, and chemical sectors; using cellulose fibres from waste to make plastic; extracting phosphate from sewage water to make struvite; making filter paper from wastewater; and using biocomposites to make a variety of products including furniture. Water utilities have installed solar panels on facility-owned infrastructure including large-scale solar arrays at a wastewater treatment plant. Several water utilities have also installed floating solar arrays on their reservoirs, with one utility using its excess renewable energy to help power public buses, the airport, health clinics, and other city services. Water utilities have installed energy storage systems, for example, the installation of an energy storage machine alongside a solar PV system at a water treatment works and the trialling of new silicon thermal energy storage technology at a wastewater treatment plant. Wind turbines too have been installed with one utility seeking new sites for future wind turbines. Finally, a water utility's drinking water is being harnessed to generate hydroelectric power.

Overall, the development of the circular water economy, which mitigates greenhouse gas emissions and enhances resilience to climate change, can be guided by innovative policies that encourage the reducing of water consumption, reusing and recycling of water, and recovery of materials from wastewater.

INDEX